Climate and plant distribution

CAMBRIDGE STUDIES IN ECOLOGY

ALSO IN THE SERIES

H. G. Gauch, Jr *Multivariate Analysis in Community Ecology*
Robert Henry Peters *The Ecological Implications of Body Size*
C. S. Reynolds *The Ecology of Freshwater Phytoplankton*
K. A. Kershaw *The Physiological Ecology of Lichens*
Robert P. McIntosh *The Background of Ecology*
Andrew J. Beattie *The Evolutionary Ecology of Ant–Plant Interactions*

Climate and plant distribution

F. I. WOODWARD
Lecturer in Botany and Fellow of Trinity Hall, University of Cambridge

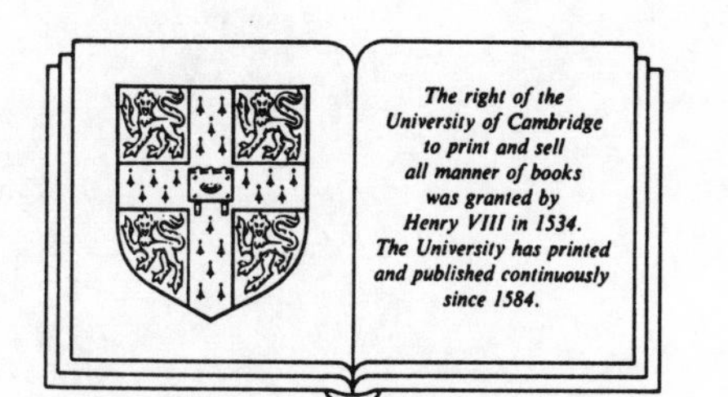

CAMBRIDGE UNIVERSITY PRESS
Cambridge
New York Port Chester
Melbourne Sydney

Published by the Press Syndicate of the University of Cambridge
The Pitt Building, Trumpington Street, Cambridge CB2 1RP
40 West 20th Street, New York, NY 10011, USA
10 Stamford Road, Oakleigh, Melbourne 3166, Australia

First published 1987
Reprinted 1988, 1990

Printed in Great Britain by the University Press, Cambridge

British Library cataloguing in publication data

Woodward, F. I.
Climate and plant distribution.
(Cambridge studies in ecology)
1. Vegetation and climate
I. Title
581.5′222 QK754.5

Library of Congress cataloguing in publication data

Woodward, F. K.
Climate and plant distribution
(Cambridge studies in ecology)
Includes bibliographies and index.
1. Vegetation and climate. 2. Phytogeography.
I. Title. II. Series.
QK754.5.W66 1987 581.5′222 86-6827

ISBN 0 521 23766 1 hardback
ISBN 0 521 28214 4 paperback

UP

To
Pearl, Helen and David

Contents

Preface

The central thesis for plant ecology is that climate exerts the dominant control on the distribution of the major vegetation types of the world. Within a vegetation type smaller-scale variations in distribution may be controlled by smaller-scale features of the environment such as soil types, human activity or topography. However, at all spatial scales the response of the plant to climate is a crucial feature in its presence.

In view of the importance of climate in controlling distribution of plants it is surprising that this area of subject is not a popular one in plant ecology. Such questions as 'when' and 'how' does climate have its effect are clearly difficult to answer, perhaps explaining the limited acceptance as an area of study. I have set out to encourage an interest in climate and plant distribution by proposing different approaches to the subject. They are approaches which may not find general acceptance, but if they create interest and debate, which is my hope, then they will have been successful.

I have drawn exclusively on vascular plants as examples of distribution types and responses. This reflects my interests and is not a comment on the ecology of non-vascular species. Where possible, I have also avoided the use of the emotive terminologies, such as niche, strategy and adaptation; however I admit to the unavoidable use of competition, in spite of real problems in its measurement.

F. I. Woodward　　　　Cambridge, 1986

Acknowledgements

This book is very much a product of the interests which I have been encouraged to develop in different academic environments. I am particularly grateful to Dr S. R. J. Woodell, Professor C. D. Pigott, Dr J. E. Sheehy, Dr J. R. Etherington, Dr M. S. Davies and Dr P. J. Grubb who have contributed in their ways to this development.

The ideas in this book have been points of discussion for some time. I am grateful for these discussions and in some cases comments on the manuscripts, by Dr G. P. Aylett, Dr H. Bauer, Miss R. C. Crabtree, Dr Q. C. Cronk, Professor J. J. Ewel, Professor R. T. T. Forman, Professor F. Franks, Dr W. Hadfield, Dr J. Harbinson, Professor C. J. Heusser, Dr V. Kapos, Dr Ch. Körner, Dr H. J. Lamb, Professor W. Larcher, Dr D. Lee, Dr W. G. Lee, Dr N. J. Shackleton, Dr W. K. Smith, Dr E. V. J. Tanner, Dr T. Webb III and Dr B. G. Williams.

The mental freedom which I have needed has been generously given by my family. I am grateful to my wife Pearl for converting my nested scribble and scrawl into a typed manuscript.

F.I.W.

Acknowledgements

1

History and demonstration

I never think of the future. It comes soon enough.
A. Einstein.

Introduction

It is intended that the title of this book should make the grand assumption that climate controls, or exerts a strong controlling influence on, the geographical distribution of plants. Any cursory glance at vegetation and climatic maps of the world cannot fail but to impress by the very close correspondence between the two distributions. In general terms, climate deteriorates from the equator to the poles and in a manner which is broadly latitudinal. Vegetation is closely analogous, so much so that general descriptions of vegetation zones such as boreal, temperate and rain forest embody a climatic connotation. However, the mechanisms which connect climate and vegetation are poorly understood. The broad objective of this volume is to investigate these mechanisms. This chapter is concerned with the development of the notion, and subsequent demonstration, that variations in climate can influence the constitution of vegetation, and therefore control plant distribution.

Theory

The awareness of this relationship in the literature dates from the peak of Ionian philosophy, between the fifth and third centuries BC. The first, very limited records on the subject are those in the fifth century BC by Menestor (Morton, 1981), who was clearly aware of the correlation between the evergreen and deciduous habits of vegetation and climate. The first extensive work on the subject of plant geography and ecology was written by Theophrastus, who lived from about 370 BC to 285 BC and was a pupil of both Plato and Aristotle. His work, *Enquiry into Plants and Minor Works on Odours and Weather Signs*, has been almost completely translated by Hort (1916) and provides a clear image of the acceptably modern ideas on plant geography held by Theophrastus. Theophrastus understood the importance of climate to plant distribution not only by observation, e.g. ' ...[that] each tree seeks an appropriate position and climate is plain from the fact that some districts bear some trees but not others ...', but also by experiment it may be seen that '... the latter do not grow there of their own accord, nor can they easily be made to grow ...[or] bear fruit ...'. This technique of transplanting individuals from their native habitats to areas outside the natural range is still one of the major techniques of experimentation in this subject.

Theophrastus was also keenly aware of the 'sympathetic relationships' between the life-cycles of plants and the season. This view was enlarged in his discussions on the 'individual properties' of plants which are directed towards survival and growth.

In any terms the work of Theophrastus was exceptional and demonstrated the highly developed philosophy of life with which Theophrastus worked. However, then as ever, the political climate of the time influenced the ideas on more sociological issues. In particular he unfortunately abandoned Anaximander's (611 – 547 BC) notion of evolution leading to the survival of better adapted forms (Morton, 1981), in favour of the socially operative, hierarchical structure of kingdoms.

Political and religious principles quashed significant developments in scientific thought, including plant geography, throughout the Middle Ages; consequently, the ideas of Theophrastus were not further developed, even into the eighteenth century.

Significant developments may be seen in the works of Willdenow in 1792 (*Grundriss der Krauterkunde*) and von Humboldt in 1807 (*Essai sur la Geographie des Plantes*), where a realisation that both climate and vegetation have, in the past, changed slowly – with the best evidence for cause and effect provided by fossil remains. The significance of fossil remains as indicators of earlier types of vegetation was, however, known to Theophrastus and the earlier Ionian philosophers. The significance of fossils to the concept of evolution was not overtly stated until Buffon, in the period of 1749 to 1767 (*Histoire Naturelle*), suggested that changes in climate can lead to the evolution and extinction of biological organisms.

The importance of both plant physiology and past history to plant geography was clearly stated by de Candolle in 1855 in his *Geographie Botanique Raisonnée*. He considered that the principal aim of plant geography was ' ...to show what, in the present distribution of plants, may be explained by present climatic conditions'. This philosophy and de Candolle's concern with the geographical distribution of particular species is little different from the philosophy of present-day plant geography, lacking only a knowledge of natural selection and Mendelian genetics and their significance to plant geography.

The fourth dimension: time

Buffon demonstrated that the fossil record could provide clear evidence for biological evolution, for historical change in plant distribution and vegetation and for the suggestion of climatic changes over millions of years.

The ability to go beyond just the suggestion of climatic control requires a knowledge of the precise time, or period, of vegetational change and an independent, concurrent measurement of climate. Determining the age of the fossils and the past climate has only been possible to any unequivocal extent in recent times. These techniques of study, using isotopic analysis, will be described later in this chapter.

A technique for demonstrating the control and mechanism by which climate could be active in controlling plant distribution, might be achieved by observations and experiments over short time scales such as a few years or decades. The emphasis on a number of years implies that, even for individual plants or species, the response time of geographical spread is very slow. For example Albertson & Tomanek (1965) studied the population dynamics of two dominant perennial grasses in Kansas, USA for nearly 30 years. They observed that the dominance of the two species was greatly influenced by drought. However the local, geographical ranges of the two species remained more or less unchanged.

There is much evidence to show that extremes of climate such as drought, low and high temperatures and high winds may influence local distribution. However there do not appear to be any direct observations on the climatic control of the geographical range or boundary of a species based on field observations at the present day. Even if observations of this type were available for individual species, it does not follow that the geographical range of whole communities of plants will change as a direct result of climate (Miles, 1979). Experiments over recent years have described the potential for the climatic control of distribution (Chapter 5) but have not observed the natural process. One of the major reasons for this lack of any real demonstration may be the rather uniform conditions observed over the period of instrumental records of about 300 years (Manley, 1974).

Geological evidence shows that the World was under the influence of an Ice Age until about 10 000 years ago and it is clear from the fossil evidence that the vegetation has changed markedly from the end of that time to the present. If the time, climate and vegetation could be simultaneously described for this period, then it should provide clear evidence for the change of geographical distribution under the control of climate.

Palaeoecology

Blytt (1876) observed plant remains – macrofossils – in anaerobic peat deposits in Denmark. The macrofossils were of parts of tree species, such as *Populus tremula* (aspen), *Pinus sylvestris* (Scots pine) and *Quercus petraea* (sessile oak). He observed that these species have rather different geographical ranges at the present day and postulated that the climate and the vegetation of Denmark had changed markedly in the past.

A critical feature of such an hypothesis is the view that the recognisable fossil species have not evolved in the period of time from the last Ice Age to the present day. In this respect it is clearly an advantage to study the recent past; however many examples of ecotypic variation have shown that both morphological and ecological evolution may occur quite rapidly, in decades, or even less. It is possible therefore, but not demonstrable, that the ecophysiological responses of plants were different in the past. If the plant communities had different component species, then it is also possible that competitive responses may also have been different. Although it is not possible to know if this is true, it is reasonable to suggest that ecotypic variation leads only to a rather fine tuning of the gross ecophysiological responses of plant species. If this is broadly true, then it is possible to accept the gross features from palaeoecology at face value.

Two problems arise in the ecological interpretation of the macrofossil remains: first, that of defining the date of the remains; second, that of predicting the composition of the vegetation from a rather poor fossil record.

Initially it was not possible to date the fossils by any absolute dating technique. Fossils from different geographical locations could be compared, in a relative way, by stratigraphical sequencing. However, even if the sequence may be aged from annual sediments (varves) it does not necessarily follow that the fossil in the sequence is of the same age.

The potential for dating fossil remains originates from the work of Libby (1946) on the radioisotopes of carbon. Atmospheric carbon occurs in a number of isotopic forms with the atomic numbers of 12, 13 or 14 (or ^{12}C, ^{13}C and ^{14}C). The stable isotope is ^{12}C and is the major component. ^{14}C is radioactive and is derived in the upper atmosphere by the interactions of cosmic ray neutrons and nitrogen. The ^{14}C derived in this way is oxidised to $^{14}CO_2$ and mixes with the more abundant $^{12}CO_2$. Both these isotopes and ^{13}C are taken up by plants and incorporated into their structures, at an equilibrium with the partial pressures of the isotopic forms of CO_2 in the atmosphere. If the plant dies and is incorporated into a sediment, the plant material is prevented from maintaining an equilibrium with the atmosphere and the radioactive ^{14}C decays, by the emission of beta rays. The rate of decay is a constant, with a half-life of 5568 ± 30 years.

If it is assumed that the ^{14}C-to-^{12}C ratio of the present day is the same as, or similar to, that at the time of deposition, then the proportionate change from the present-day ^{14}C-to-^{12}C ratio to that in the sediment may provide an estimate of the age of the plant remains and, therefore, the age of the sediment. A number of sources of error are likely, as discussed by West (1977). Typical confidence ranges for an estimate of age are about ±10–20%, increasing with age. In addition, Ralph & Michael (1967) show a systematic deviation between ^{14}C dates and dates obtained from a long series of tree rings (from *Sequoia gigantea* and *Pinus aristata*). The limit of reliable dating is about 40 000 years, although Miller (1977) has shown the potential for dating to 100 000 years using a cyclotron. It is clear that radiocarbon dating is an ideal technique for establishing chronologies alongside palaeoecological evidence in the late Quaternary.

The second problem of interpreting the ecological evidence provided by the sparse macrofossil evidence created a considerable debate in the period from Blytt's publication in 1876 to about 1916. In 1916, von Post gave a lecture on the occurrence of forest-tree pollen in peat from southern Sweden (von Post, 1918). In his introduction, von Post reviewed the evidence from macrofossils for historical changes in vegetation, and suggested that their fortuitous occurrence provided, in general, an incorrect picture of the chronology of tree immigration. Von Post then provided evidence for the changing geographical distributions of tree species in Sweden since the last Ice Age. His technique was that of the identification and counting of pollen grains within local stratigraphical sequences.

This notion of palynology as a palaeoecological tool was probably conceived in a qualitative manner by Geinitz (1887) and Weber (1893), (Faegri & Iversen, 1975). However von Post was the first to use the information quantitatively. He was quick to realise the full potential of palynology, stemming from the greater abundance of pollen grains in sediment than macrofossils (1000 times greater, or more), the persistence of the resistant pollen grains in anaerobic conditions and their approximate correlation with the abundance of the parent plants over quite large geographical areas (Birks & Birks, 1980). Although the pollen grains of a plant are a constant taxonomic feature, their identification to the specific level is less often realised and, in this respect, they are therefore less informative than many macrofossils.

The work of von Post provided a clear description of the way in which the arboreal vegetation of Sweden has changed since the last Ice Age. Some of his data are presented in Fig. 1.1 (from Fries, 1967). The

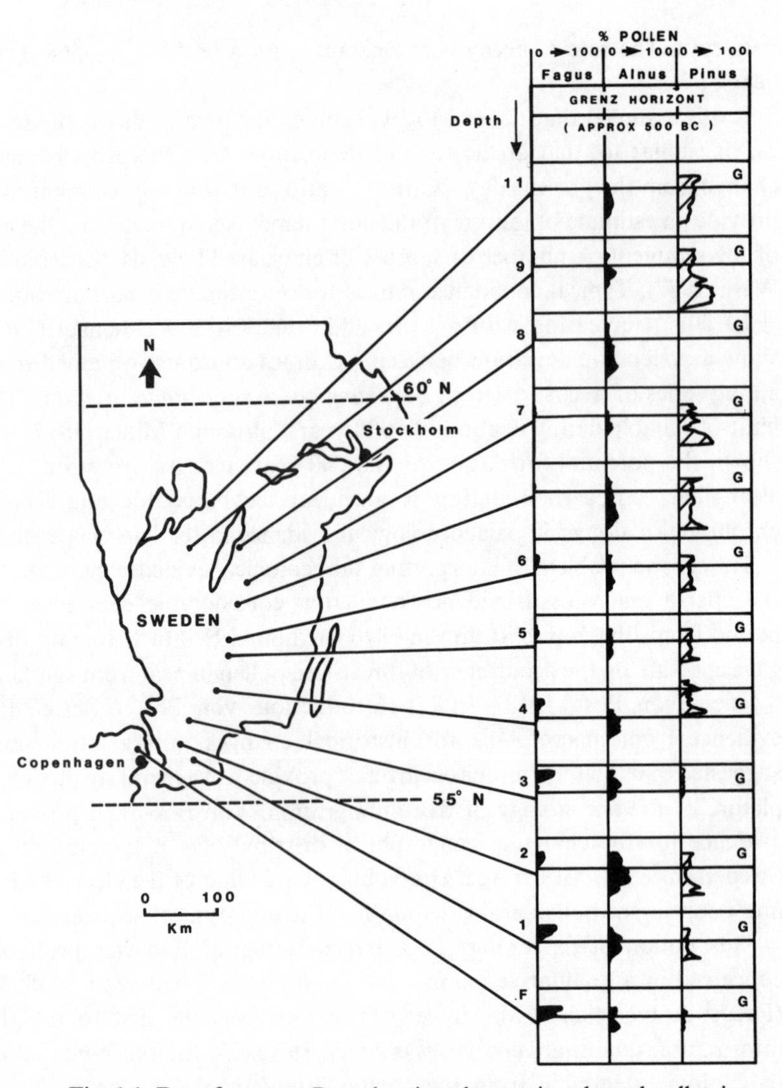

Fig. 1.1. Data from von Post showing changes in arboreal pollen in Sweden.

percentage of the sampled pollen due to one species is represented as a histogram (abscissa) with the depth of the core of peat on the ordinate.

At the time of its publication, the pollen diagrams could not be dated by ^{14}C techniques. However a recognisable stratigraphic horizon, the 'Grenzhorizont', could be identified in the peat cores. The Grenzhorizont separates humified from poorly humified peat and has been dated at about 500 BC, although there is evidence that the age of this horizon may vary within and between sites and perhaps with more than one boundary

(Faegri & Iversen, 1975). In spite of this uncertainty it is clear from Fig. 1.1 that *Fagus*, a species found predominantly in the warmer south of the country, had only been present since the Grenzhorizont and then mainly in the south. *Pinus*, a species of cooler climates, on the other hand, was evidently abundant before this time and has declined since the horizon. The interval between the times of abundance of *Pinus* and *Fagus* was characterised by an increase in *Alnus*, suggesting a wet and warm climate.

Von Post also viewed his results in a dynamic manner, intimating that species spread northwards from the warmer south during the amelioration of climate after the last Ice Age, and that this spread was likely to have been rapid.

Considerable developments in palaeoecology have occurred since the time of von Post's early paper. A particular concern has been the relationship between plant frequency and pollen production. Genera with entomophilous pollination, such as *Tilia*, produce less pollen than anemophilous genera, such as *Betula* and *Pinus*, and the latter are therefore overrepresented in pollen diagrams. Davis (1963) introduced an empirical correction factor to account for these differences. The aim of these methods is an improvement in the predictive capabilities of palaeoecology. However, in the search for incontrovertible evidence of the climatic control of plant distribution, these techniques are overshadowed by the development of another isotopic technique, using the isotopes of oxygen.

Palaeothermometry

It is probably reasonable to suppose that the occurrence of, for example, *Fagus* in Fig. 1.1 indicates a moist, temperate climate. However this argument is circular and provides no measurement of climate which is independent of the fossil record. West (1964) has observed different combinations of tree taxa during previous interglacial periods in Europe. He concluded that the specific composition of the vegetation of the present has no antecedent in the Quaternary, and is merely a temporary aggregation. It can be unwise, therefore, to predict the climate of the past on the basis of the occurrence of pollen.

Emiliani (1955) was probably the first to appreciate the importance to Quaternary palaeoecology of Urey's observations (1947) on the differential fractionation of the isotopes of oxygen, ^{18}O and ^{16}O, as an indicator of the relative or even absolute temperatures through a chronological sequence. The isotope ^{16}O is considerably more common (99.76%) than ^{18}O (0.2%) in the atmosphere. These two isotopes are found in water as $H_2{}^{16}O$ and $H_2{}^{18}O$. Their natural occurrence varies because of differential evaporation and condensation. During evaporation from a mass of water, the water vapour is depleted in ^{18}O (about 1%, Gray, 1981) relative to the liquid phase, because of the greater vapour pressure of $H_2{}^{16}O$. Therefore the remaining liquid water becomes enriched in ^{18}O. During condensation the relative frequencies of the two isotopes also change and ^{18}O (as $H_2{}^{18}O$) will increase in the condensate. If condensation occurs immediately after evaporation, the condensed water should have the same isotopic ratio as the mass of water. Cooling the air will increase the rate of condensation and lead to a decrease in the ^{18}O component of the vapour phase. The ^{18}O-to-^{16}O ratio of rainfall will therefore decrease monotonically with temperature (Dansgaard, 1964). If this precipitation is then incorporated into terrestrial materials and isolated from subsequent isotopic contamination, then a measure of the isotopic ratio will provide a measure of the temperature at which condensation occurred and, therefore, of air temperature. Suitable terrestrial materials for analysis include deposits of calcium carbonate, including the shells of organisms such as foraminiferans, snow or ice, peat and the α-cellulose component of the wood in trees (Gray, 1981).

The two isotopes of oxygen in the sample are measured as a ratio to a standard. Two standards are usually employed, either standard mean ocean water (SMOW) or Pee Dee belemnites (PDB) (Gray, 1981). The ^{18}O-to-^{16}O ratio ($\delta^{18}O$) is positive in value when the $\delta^{18}O$ of the sample exceeds the reference, and negative if it is less than the reference. The concentration of ^{18}O is small and so the ratio of ^{18}O-to-^{16}O is per thousand (mil).

In ideal circumstances, with no contamination, the temperature coefficient is 0.7‰ $°C^{-1}$ (Dansgaard, 1964) and the best expected resolution could be 0.1 °C. However when the isotopic ratio of biological material and sedimentary deposits is considered, resolution will be degraded to perhaps 0.5 °C at best (Gray, 1981) and absolute temperature equivalents may not be possible. This problem arises because the material, with incorporated rainfall and a certain $\delta^{18}O$, will be deposited in liquid water, perhaps differing in its ^{18}O-to-^{16}O ratio from that of the rainfall.

The source of much of the global precipitation may be considered to be the tropical oceans (Hendy & Wilson, 1968). If the $\delta^{18}O$ of these oceans changed, which is almost certain to be the case during the ice ages when the mean global temperature will have declined, decreasing the $\delta^{18}O$ in rainfall and increasing ^{18}O in sea water, with the isotopically light water being stored in glacial ice (Shackleton & Opdyke, 1973), then the predicted temperature from $\delta^{18}O$ will be in error.

Evidence for the significance of the glacial control of $\delta^{18}O$ has been achieved by analysing the isotopic ratio of surface living (pelagic) foraminifers and those of deep water living (benthic) foraminifers. The pelagic foraminifers should show sensitivities in $\delta^{18}O$ due to both temperature and glacial effects, while the benthic foraminifers should be buffered from the effects of changing temperature but not glacial extent (Dansgaard & Tauber, 1969). Dansgaard & Tauber (1969) conclude from the evidence provided by foraminifers and from glaciological evidence on the latitudinal extent of the polar ice caps (Flint, 1971), that only about 30% of the variation in $\delta^{18}O$ observed in cores results from changes in temperature, while 70% is because of the glacial control of the isotopic composition of sea water, i.e. the extent of the ice caps.

Correlations between climatic and vegetational change

The complex background to the use of oxygen isotopes as palaeothermometers is a necessary prelude to their use as indicators of climatic change. All the published details on cores which have been analysed for $\delta^{18}O$ discuss the likely errors which might prevent an interpretation in terms of temperature alone. The majority of these cores have been taken from the oceans and ice caps of the World. However, in the context of correlating changes in terrestrial vegetation with changes in climate measured by temperature, the necessary information must come from parallel and terrestrial observations on both the isotopic ratios and reconstructions of past vegetation.

Three analyses of $\delta^{18}O$ have been selected for consideration, and have been measured by Hendy & Wilson (1968), Stuiver (1968) and Eicher, Siegenthaler & Wegmuller (1981). Stuiver and Eicher *et al.* carried out their measurements of $\delta^{18}O$ on marl deposited in sediments under lakes (lacustrine). Lacustrine marl is rich in carbonates and is mainly derived from aquatic photosynthesis, which removes carbon dioxide dissolved in the lake water and leads to the conversion of soluble bicarbonates to insoluble calcium carbonate, which is therefore precipitated. This reaction occurs principally in the warm season and therefore can provide a measure of summer temperatures. The $\delta^{18}O$ of the carbonate reflects the $\delta^{18}O$ and therefore temperature, of the surrounding solution. However, the $\delta^{18}O$ for the formation of carbonate has a negative temperature coefficient of $-0.25‰\ °C^{-1}$ (Eicher *et al.*, 1981), while the $\delta^{18}O$ of the water supplied by rain has a positive temperature coefficient of $0.7‰\ °C^{-1}$. These two coefficients may be combined if the effects of changes in the $\delta^{18}O$ of ocean water on precipitation are known (Stuiver, 1968).

Hendy & Wilson (1968) have determined a series of $\delta^{18}O$ values for calcium carbonate (calcite) deposited in caves (speleothems). The calcium carbonate deposits form in isotopic equilibrium with water which seeps

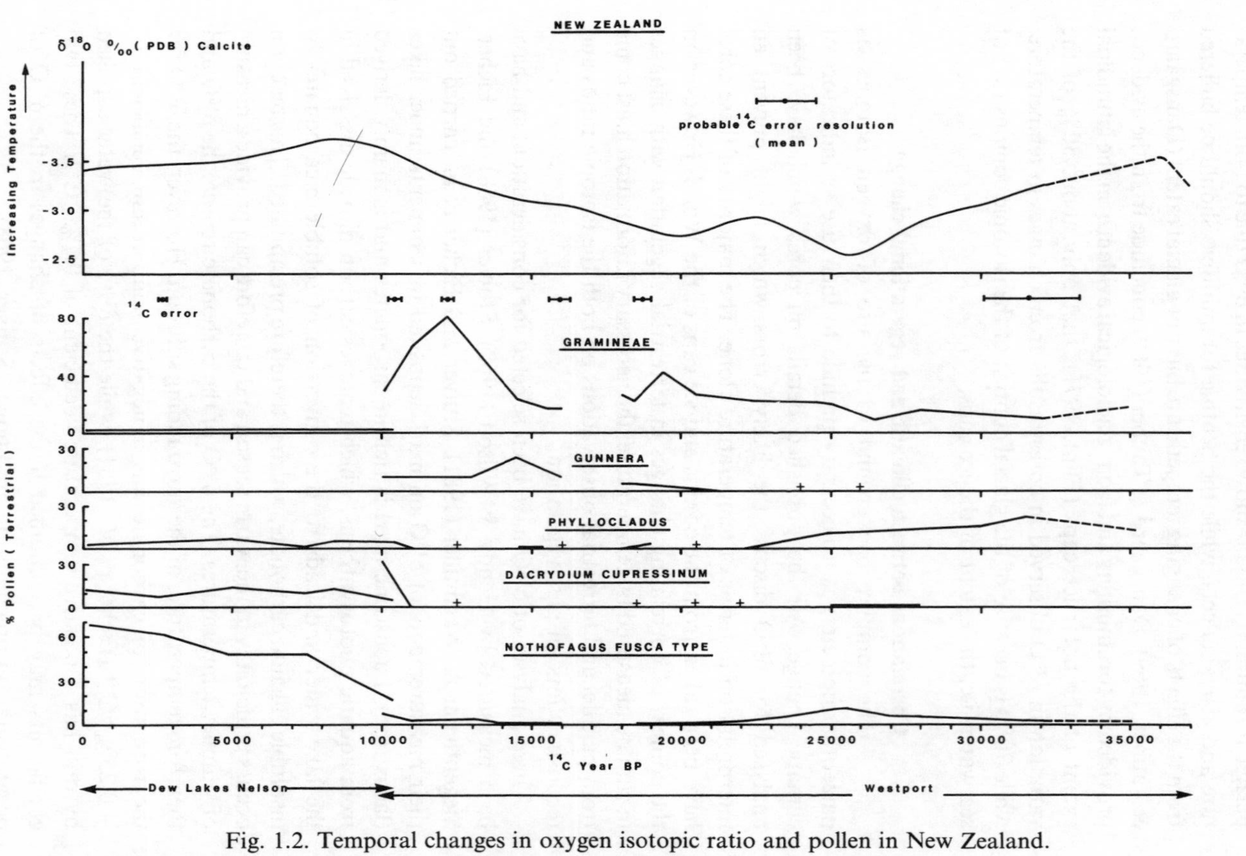

Fig. 1.2. Temporal changes in oxygen isotopic ratio and pollen in New Zealand.

into the cave. The data from speleothems suffer from the same problems of interpretation as marl because of the unknown $\delta^{18}O$ of the seepage water. However $\delta^{18}O$ will have a closer correlation with the annual temperature than the summer temperature, because of the insulating effects of large masses of rock. The $\delta^{18}O$ for speleothems has a negative temperature coefficient. Wilson, Hendy & Reynolds (1979) have demonstrated very close agreement between the mean annual temperatures predicted from isotopic measurements of speleothems and instrumental records over the last 300 years, indicating good faith in the techniques of measurement and analysis, at least in the short term.

Given this confidence in the suitability of the oxygen isotope technique as a palaeothermometer, it should be possible to use the technique alongside that of pollen analysis to provide an independent measure of climate. The aim here is simple, that of indicating a correlation between the changes in climate, measured by changes in $\delta^{18}O$, and changes in the vegetation, predicted from changes in the frequencies of pollen grains. If such correlations can be demonstrated, then it may be concluded (although not unequivocally) that climate has controlled plant distribution.

Hendy & Wilson (1968) established a chronology of about 40 000 years for their isotopic measurements of speleothems, on the North Island of New Zealand. No pollen diagrams, with chronologies established by ^{14}C-dating, have been found close to the site of the caves, at Waitomo, Te Kuiti, 38 ° S. Two dated pollen diagrams from the northern end of the South Island, at Nelson, 41 ° S (from Dodson, 1978), and Westport, 41 ° S (from Moar & Suggate, 1979) have been spliced to provide a chronology from about the present day to 31 600 years before the present (the BP datum point is AD 1950). The splice is at about 10 000 years BP, marking the end of the Nelson diagram and the start of the Westport diagram.

The pollen diagrams (for *Nothofagus fusca*, the evergreen red beech, *Phyllocladus*, a coniferous shrub, and Gramineae, representing the undifferentiable grasses) and the $\delta^{18}O$ ratios are shown on Fig. 1.2. The abscissa is the time axis, unlike the normal palaeoecological convention as shown in Fig. 1.1. The likely errors in establishing the chronology are also shown.

The trend in $\delta^{18}O$ demonstrates a major warm period at about 8 000 BP, declining slightly to the present day but declining markedly to temperature minima at about 20 000 BP and 26 000 BP. These periods of low temperature indicate periods of global glacial advance (in the period of Ice Age conditions). Clear correlations between climate and pollen frequencies also emerge. *Nothofagus fusca*, a temperate rain forest tree, dominates the pollen diagram, and probably the local vegetation, from about 10 000 BP, increasing initially in abundance with the increase in temperature up to 8 000 BP. The grasses dominated the vegetation during the coldest period from 30 000 to about 10 000 BP. The presence of *Phyllocladus* up to about 22 000 BP indicates that the vegetation over this period was more shrub-like, with a limited occurrence of *Nothofagus*. *Phyllocladus* has also been present in the vegetation from 11 000 BP, probably as an understorey or secondary component of the *Nothofagus* forests.

In general terms the data on Fig. 1.2. do indeed suggest that changes in climate are correlated with the local vegetation, with grasses dominating in the coldest period, a shrubby vegetation in the cool periods and a *Nothofagus* forest in the warmest period from about 9 000 BP. However, it is also interesting to note that similar percentages of pollen in the diagram, e.g. *Nothofagus* at 25 000 and 10 000 BP, and *Phyllocladus* at 24 000 and 5 000 BP do not correlate with the same value of $\delta^{18}O$. Similarly, the same values for $\delta^{18}O$ at 14 000 and 31 600 BP do not correspond with equal occurrences of specific pollen. These differences may reflect the different response times of species to the changes in climate, and is evidence for the view of West (1964) that plant communities are merely temporary aggregations (see also Chapter 2).

The second analysis (Fig. 1.3) combines the oxygen isotope analyses of Stuiver (1968), at Pretty Lake, Indiana, in the USA (41° N), with a pollen diagram from the same lake (Ogden, 1969). The chronologies of both series have been established by ^{14}C dating and are much shorter than those for the series in New Zealand.

The oxygen isotope ratios are broadly similar to the pattern observed in New Zealand, with a decline in $\delta^{18}O$, perhaps representing a fall in mean temperature of 6–7 °C, from a peak at 8 000–9 000 BP. The more rapid decline in Indiana may reflect real differences between the northern and southern hemispheres of the world or may, for example, be the result of the lower frequency of sampling in New Zealand.

The percentage frequency of pollen of three arboreal genera *Pinus*, *Quercus* and *Ulmus* have been included for comparison. Between 10 000 and 11 000 BP the boreal genus, *Pinus*, dominated the community. However this period was short-lived and the climatic amelioration to the peak at 8 000–9 000 BP was accompanied by a large decline in *Pinus* and an increase in *Quercus*, which must have immigrated into and overwhelmed the existing pine forests, clearly with some time lag in the change of the vegetation. The changes in climate reflected in the $\delta^{18}O$ were less extreme from 8 000 BP to 2 700 BP and *Quercus* remained dominant throughout the period. The peak in $\delta^{18}O$ at 5 600 BP corresponds with an increase in the

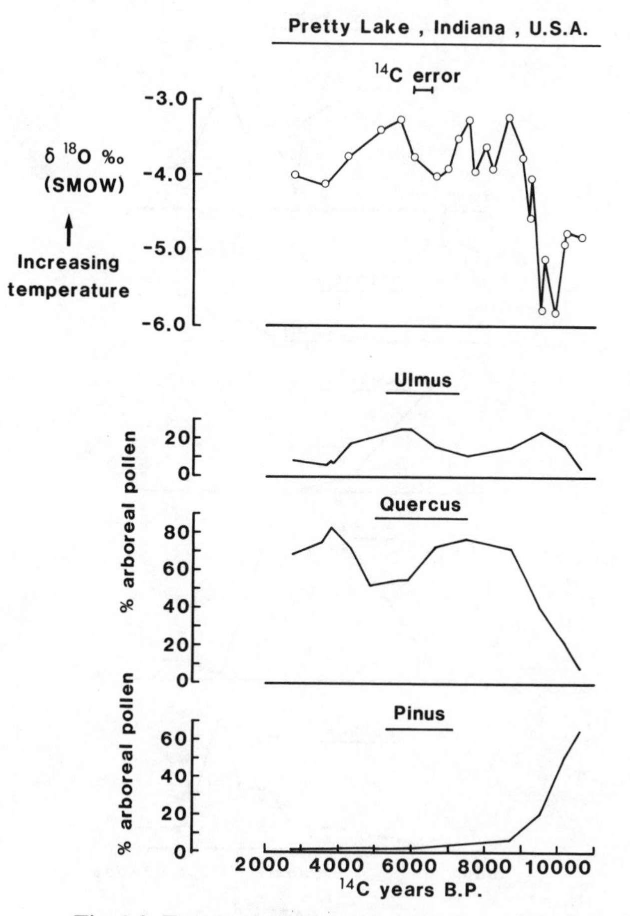

Fig. 1.3. Temporal changes in oxygen isotopic ratio and pollen at Pretty Lake, Indiana, U.S.A.

pollen of the more thermophilic *Ulmus*, at the expense of *Quercus*. *Ulmus* also reached a peak in occurrence at 9600 BP during the increase in *Quercus*. This peak corresponds with another but lower peak in $\delta^{18}O$.

Part of the work of Eicher *et al.* (1981) is presented in Fig 1.4. This work was a combined palynological and isotopic analysis of a core taken from a bog in Dauphine, France, (45 ° N). Unfortunately it was only possible to establish a chronology by stratigraphic techniques with one time-marker, at 11 000 BP. However the stratigraphic and palynological series correlate quite well with other dated sequences in Europe and so it has been possible to derive an approximate chronology.

Pollen records for the arboreal genera *Betula*, *Pinus*, *Corylus* and the herbaceous genus of *Artemisia* are included. The oldest sediments

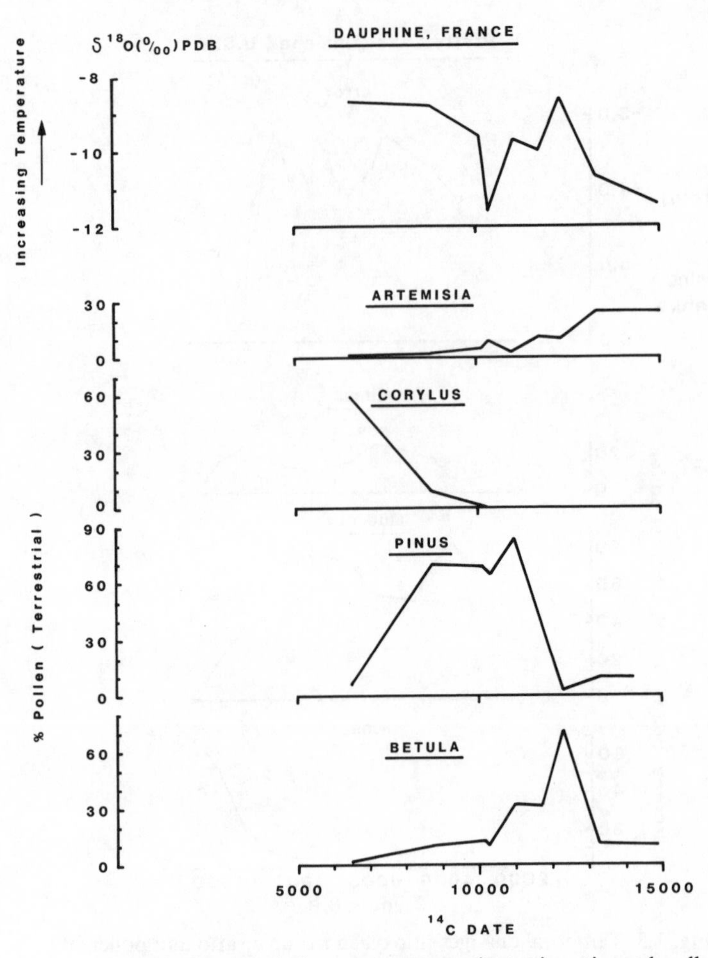

Fig. 1.4. Temporal changes in oxygen isotopic ratio and pollen at Dauphine, France.

shown in Fig. 1.4. were derived at about 13 000–15 000 BP and were characterised by a low $\delta^{18}O$, indicating low temperatures. Pollen of *Artemisia* was plentiful and characterised a late-glacial flora. The increase in $\delta^{18}O$, perhaps representing an increase in mean temperature of 7 °C to about 12 300 BP, correlates precisely with an increase in the pollen of *Betula*. It is not possible to differentiate between the pollen of the dwarf birch, *Betula nana*, and the tree birches, such as *B. pubescens*. However, other shrubby species from 13 000 BP and earlier were also reduced at this time (Eicher *et al.*, 1981), suggesting a prevalence of tree birch.

The $\delta^{18}O$ decreased from a peak of 12 300 BP to a minimum at about 10 500 BP, correlating quite well with the observation in Indiana (Fig. 1.3).

During this period, the pollen of the boreal genus *Pinus* increased markedly and rapidly, suggesting its replacement of *Betula* and the formation of pine forests. The low $\delta^{18}O$ at 10 500 BP saw a small decline in both *Betula* and *Pinus* and an increase in *Artemisia*, perhaps indicating a more open system of pine woodlands rather than dense forests.

The $\delta^{18}O$ increased again following the minimum, after a rather short period of low temperatures, a feature also found in Indiana (Fig. 1.3). During the period of increasing $\delta^{18}O$, and therefore temperature, to about 6500 BP first *Betula* and then *Pinus* decreased, with an increasing proportion of the temperate genus, *Corylus*.

Conclusions

The evidence presented in Figs. 1.2 – 1.4 clearly suggests that both climate and vegetation have changed over the last 32 000 years and that these changes were probably worldwide. The correlation between the occurrences of plant species, as indicated by the pollen record, and changes in climate, as determined by $\delta^{18}O$, suggests that climate was the controlling factor. Further evidence for climate control is provided by the synchronous changes in the vegetation during the period of about 13 000– 10 000 BP, a period of lower temperature than the present day.

The evidence in Figs. 1.2 – 1.4 is provided by samples from one point in space, at each of three locations. The data may not therefore indicate large scale changes in the vegetation, although this interpretation is unlikely because the pollen which accumulates at the sites will have travelled considerable distances (Birks & Birks, 1980). If evidence can be provided to show changes in geographical distribution over a large geographical area and to show that these changes occurred more or less synchronously with changes in climate, as measured by $\delta^{18}O$, then it is reasonable to conclude that climate has controlled plant distribution.

This evidence is provided on Fig. 1.5, which shows the change in $\delta^{18}O$ from about 10 000 BP to 2500 BP at Pretty Lake, Indiana (Stuiver, 1968) Fig. 1.5(*a*); the change in the altitude of the tree line in California (data of La Marche, presented in Lamb, 1982) from dated macrofossil remains of tree trunks Fig. 1.5(*b*); the change in altitude of the tree line in Colorado, predicted from past and present day ratios of pollen grains (Maher, 1972) Fig. 1.5(*c*); the changes in the alpine tree line in temperate latitudes of the world and the arctic tree line in North America and northern Europe, determined from macrofossil remains (Markgraf, 1974) Fig. 1.5(*d*).

It is clear from Fig. 1.5. that the tree lines fell during the cold period at about 10 000 BP and rose, particularly in the warm period, from about 6000–5000 BP. The palaeoecological evidence clearly demonstrates the

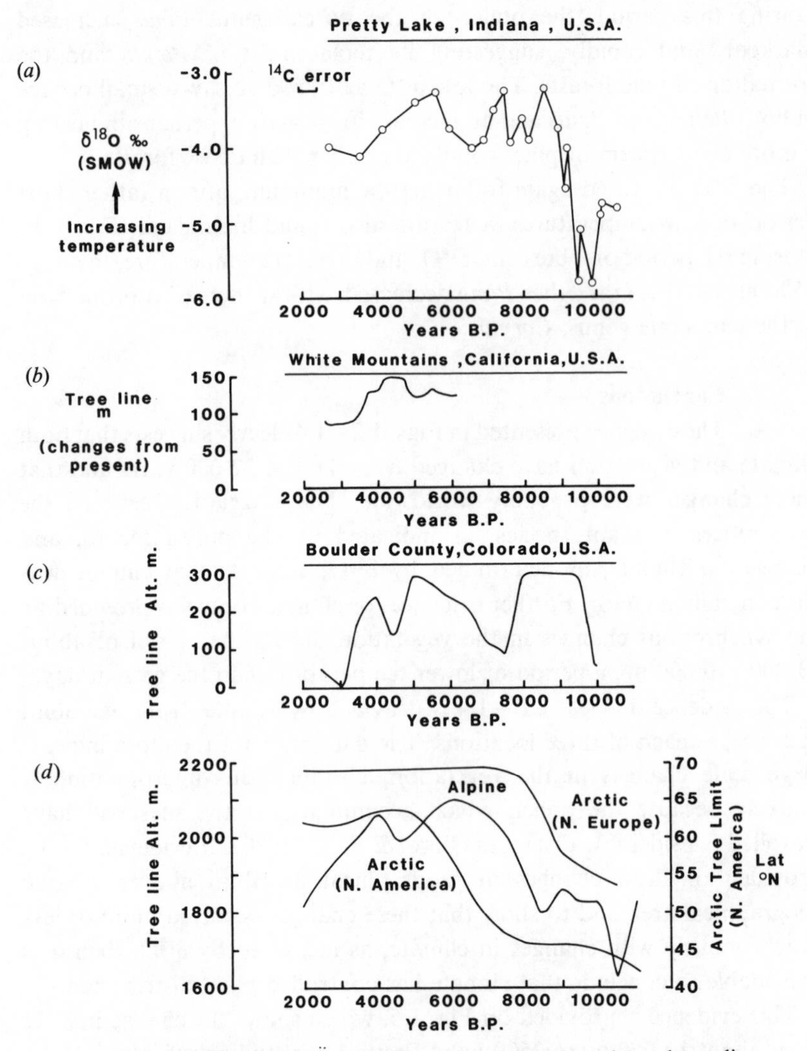

Fig. 1.5. Temporal changes in oxygen isotopic ratio and tree lines.

dynamic nature of both climate and vegetation, with continuous change being the rule rather than the exception. Precise chronological agreement between the measurements of $\delta^{18}O$ and the biological measurements does not occur throughout and may be due to both lag in the response of the vegetation and also to errors in dating. However, this does not detract from the very clear demonstration of the climatic correlation with plant distribution. The demonstration, of course, implies nothing more: the mechanism of the control, which may at one extreme be physiological or at the other pedological, is not known. The analysis of the mechanisms of control is precisely suited to observations of the present day and is the concern of the remainder of this volume.

References

Albertson, F.W. & Tomanek, G.W. (1965). Vegetation changes during a 30 – year period in grassland communities near Hays, Kansas. *Ecology*, **46**, 714–20.

Birks, H.J.B. & Birks, H.H. (1980). *Quaternary Palaeoecology*. London: E. Arnold.

Blytt, A. (1876). *Essay on the Immigration of the Norwegian Flora during Alternating Rainy and Dry Periods*. Christiana: Cammermyer.

Dansgaard, W. (1964). Stable isotopes in precipitation. *Tellus* **16**, 436–68.

Dansgaard, W. & Tauber, H. (1969). Glacier oxygen-18 content and Pleistocene ocean temperatures. *Science*, **166**, 499–502.

Davis, M.B. (1963). On the theory of pollen analysis. *American Journal of Science*, **261**, 897–912.

Dodson, J.R. (1978). A vegetation history from north-east Nelson, New Zealand. *New Zealand Journal of Botany*, **16**, 371–8.

Eicher, U., Siegenthaler, U. & Wegmuller, S. (1981). Pollen and oxygen isotope analyses on late- and post-glacial sediments of the Tourbière de Chirens (Dauphine, France). *Quaternary Research*, **15**, 160–70.

Emiliani, C (1955). Pleistocene temperatures. *Journal of Geology*, **63**, 538–78

Faegri, K. & Iversen, J. (1975). *Textbook of Pollen Analysis*. 3rd edn. Oxford: Blackwell Scientific Publications.

Flint, R.F. (1971). *Glacial and Quaternary Geology*. New York: Wiley.

Fries, M. (1967). Lennart von Post's pollen diagram series of 1916. *Review of Palaeobotany & Palynology*, **4**, 9–13.

Gray, J. (1981). The use of stable-isotope data in climate reconstruction. In *Climate and History: Studies in Past Climate and their Impact on Man*, ed. Wigley, T.M.C., Ingram, M.J. & Farmer, G., pp. 53–61. Cambridge University Press.

Hendy, C.H. & Wilson, A.T. (1968). Palaeoclimatic data from speleothems. *Nature*, **219**, 48–51.

Hort, A. (1916). *Enquiry into Plants and Minor Works on Odours and Weather Signs*. By Theophrastus and translated by Sir Arthur Hort, vol. I & II. London: W. Heinemann.

Lamb, H.H. (1982). *Climate History and the Modern World*. London: Methuen.

Libby, W.F. (1946). Atmospheric helium three and radiocarbon from cosmic radiation. *Physical Review*, **69**, 671–2.

Maher, L.J. (1972). Absolute pollen diagram of Redrock Lake, Boulder County, Colorado. *Quaternary Research*, **2**, 531–53.

Manley, G. (1974). Central England temperatures: monthly means 1659 to 1973. *Quarterly Journal of the Royal Meteorological Society*, **100**, 389–405.

Markgraf, V. (1974). Paleoclimate evidence derived from timberline fluctuations. In *Colloques Internationaux du C.N.R.S.*, No. 219, ed. I. Labeyrie, Les méthodes quantitives d'étude des variations du climat au cours du Pléistocène, pp. 67–77. Paris: Centre de la Recherche Scientifique.

Miles, J. (1979). Vegetation dynamics. *Outline Studies in Ecology*. London: Chapman & Hall.

Miller, R.A. (1977). Radioisotope dating with a cyclotron. *Science*, **196**, 489–94.

Moar, N.T. & Suggate, R.P. (1979). Contributions to the Quaternary history of the New

Zealand flora. 8. Interglacial and glacial vegetation in the Westport District, South Island. *New Zealand Journal of Botany*, **17**, 361–87.

Morton, A.G. (1981). *History of Botanical Science*. London: Academic Press, 474 p.

Ogden, J.G. III. (1969). Correlation of contemporary and late Pleistocene pollen records in the reconstruction of postglacial environments in North Eastern North America. *Mitteilungen der Internationalen Vereinigung für theoretische und angewandte Limnologie*, **17**, 64–77.

Post, L., von (1918). Skogsträdpollen i sydvenska torvomosselagerföljder. *Förhandlinger ved de Skandanaviske naturforskeres möte* (1916), 432–65.

Ralph, E.K. & Michael, H.N. (1967). Problems of the radiocarbon calendar. *Archaeometry*, **10**, 3–11.

Shackleton, N.J. & Opdyke, N.D. (1973). Oxygen isotope and palaeomagnetic stratigraphy of equatorial Pacific core v28–238: Oxygen isotope temperatures and ice volumes on a 10^5 and 10^6 year scale. *Quaternary Research*, **3**, 39–55.

Stuiver, M. (1968). Oxygen-18 content of atmospheric precipitation during last 11,000 years in the Great Lakes Region. *Science*, **162**, 994–7.

Urey, H.C. (1947). The thermodynamic properties of isotopic substances. *Journal of the Chemical Society*, **152**, 190–219.

West, R.G. (1964). Inter-relations of ecology and Quaternary palaeobotany. *Journal of Ecology*, **52** (Suppl.), 47–57.

West, R.G. (1977). *Pleistocene Geology and Biology with Especial Reference to the British Isles*. London: Longman.

Wilson, A.T., Hendy, C.H. & Reynolds, C.P. (1979). Short-term climate change and New Zealand temperatures during the last millenium. *Nature*, **279**, 315–17.

2

Scale

The hidden soul of harmony.
J. Milton.

Introduction

The previous chapter has provided evidence for the strong correlation between the distribution of vegetation and climate. The examples came from palaeoecological evidence, with a temporal resolution little better than a few centuries. So far, interpretation of these data has not established the relevance of shorter time scales in the control of distribution. From present-day observations it appears unlikely that the minute to minute variations in, for example, temperature during average weather conditions are directly important in this control. Yet the fluctuations of temperature on this scale may be similar, or even greater, than the month to month variations in mean temperature, which can influence the distribution or occurrence of species, particularly those with life-cycles extending over a year or less. There is a suggestion, therefore, of a close link, at least in this case, between the time for the life-cycle to be completed, defined as the developmental response time (Woodward & Sheehy, 1983), and the temporal variation of the controlling climatic process.

Response times have been used as analytical tools in evolutionary ecology (e.g. Lewontin, 1966; Levins, 1968 and Wimsatt, 1980). Lewontin, for example, has suggested that a changing environment may only influence the dynamics of natural selection if the response times of the environmental change and the growth, or generation time, of the population are closely matched. He proposed that organisms will adapt physiologically (i.e. a short response time) to those changes in the environment which are much shorter than the generation time, while selection in response to a change which is much longer than the generation time will never occur, because the organism will have 'forgotten' the last cycle of events before the next one occurs.

Levins (1970), Allen (1977) and Delcourt, Delcourt & Webb (1983) have extended this view to hierarchical systems, where an increased level of complexity, as for example in the series from cell to plant to population, would be characterised by an increase in the response time to a climatic change.

This approach could be of value to the problem presented here, that of the search for the important time scales in the control of vegetational change.

Response times

The response time or time constant, τ, can be precisely defined and most easily understood in terms of the response of an object, such as a leaf, to a step change in environmental conditions. The step change may be considered in terms of temperature, and in this case the time for a leaf to change by 63.2% towards the new temperature defines τ. Therefore:

$$\frac{dT}{dt} = \frac{T_f - T}{\tau} \tag{1}$$

where dT/dt is the rate of change of the object temperature, T_f is the final temperature and T the temperature at any instant. This equation may be solved as follows:

$$\frac{T_f - T}{T_f - T_0} = e^{-t/\tau}, \tag{2}$$

where T_0 is the initial temperature and t is the time. The leaf will reach a temperature equal to T_0 plus 95% of the difference between T_0 and T_f after 3 time constants and 99% after 4.6 time constants.

Few changes in the environment occur as step changes, with slow fluctuations being the rule. However the time constant is still valuable in two distinct ways. The end result of a finite response time is that, in the case of temperature for example, an object will lag behind the environmental change. This time lag (t_l) may be defined as follows (Woodward & Sheehy, 1983), in response to a typical diurnal and sinusoidal fluctuation of temperature:

$$t_l = \frac{\arctan(2\pi f\tau)}{2\pi f} \tag{3}$$

where f is the frequency at which temperature fluctuates.

In addition to the lag in response to an environmental change, the object will also experience a diminished derivative of the signal such as temperature. This attenuation of the environmental change, α, may also be defined in terms of the time constant:

$$\alpha = (1 + 4\pi^2 f^2 \tau^2)^{-\frac{1}{2}} \tag{4}$$

where the attenuation factor is a multiplicative coefficient.

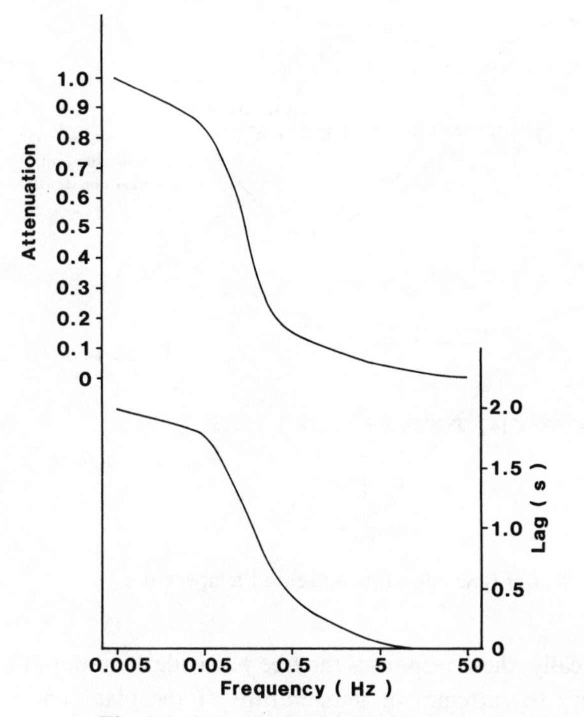

Fig. 2.1. Attentuation and lag at a range of frequencies with a time constant of 2 seconds.

The relationship between time constant, attenuation and time lag may be seen on Fig. 2.1, for a fixed time constant of 2 s, and over a range of 4 orders of magnitudes in frequency. The relationship between frequency (logarithmic scale) and both lag and attenuation is sigmoidal over the selected range. In broad terms, frequencies which are at least ten times less than the inverse of the response time, the response frequency, are perceived by an object with little attenuation and negligible lag relative to the signal. Frequencies which are ten times greater or more than the response frequency are not perceived, whilst frequencies close to the response frequencies are sensed with varying degrees of attenuation and lag.

The effects of the two time constants, 1 h, equivalent to a very thick and broad leaf, and 5 h, equivalent to a plant stem about 30 mm thick (see Woodward & Sheehy, 1983), on the temperature at the centre of the object are shown on Fig. 2.2(*a*) for a diurnal cycle of temperature. As expected from **(3)** and **(4)**, the effect of an increase in the response time is a diminution of temperature and an increasing lag behind the cycle of air

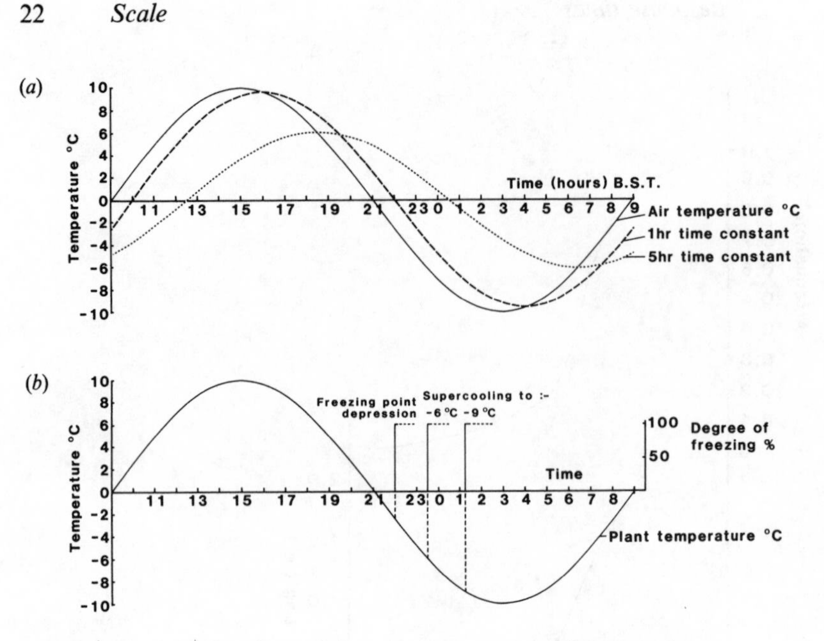

Fig. 2.2. The effect of time constants on sensed temperature.

temperature. Biologically, these responses may be particularly important in terms of sensitivity to extremes of temperature. If the plant under consideration has a freezing point of −9 °C, then this would still be sensed by the leaf but avoided by the stem which fails to cool below about −6 °C.

Freezing of plants is an all-or-nothing phenomenon in terms of survival, with intracellular freezing leading to death. A number of physical and biological mechanisms occur which serve to avoid freezing (see Chapter 4). Solutes may accumulate within the cell and serve to depress the freezing point by 2–3 °C. If this was the only method of avoidance then freezing could occur about 1 h after the plant temperature fell below the freezing point of pure water at 0 °C (Fig. 2.2(*b*)). However, water supercools below the freezing point in most tissues. The natural range of supercooling is very broad but, in the particular example of −6 °C (Fig. 2.2(*b*)), the process effectively delays the onset of freezing by 2.5 h after the plant temperature falls below 0 °C. Plants are also able to control intracellular freezing by facilitating extracellular freezing. The range of this control is again wide, but in the example a freezing resistance to −9 °C would provide a delay in freezing of 4.25 h. None of the examples provides complete protection over the diurnal cycle of temperature. However a freezing resistance of just under −10 °C would provide complete protection and provide a selective advantage for individuals with this capacity.

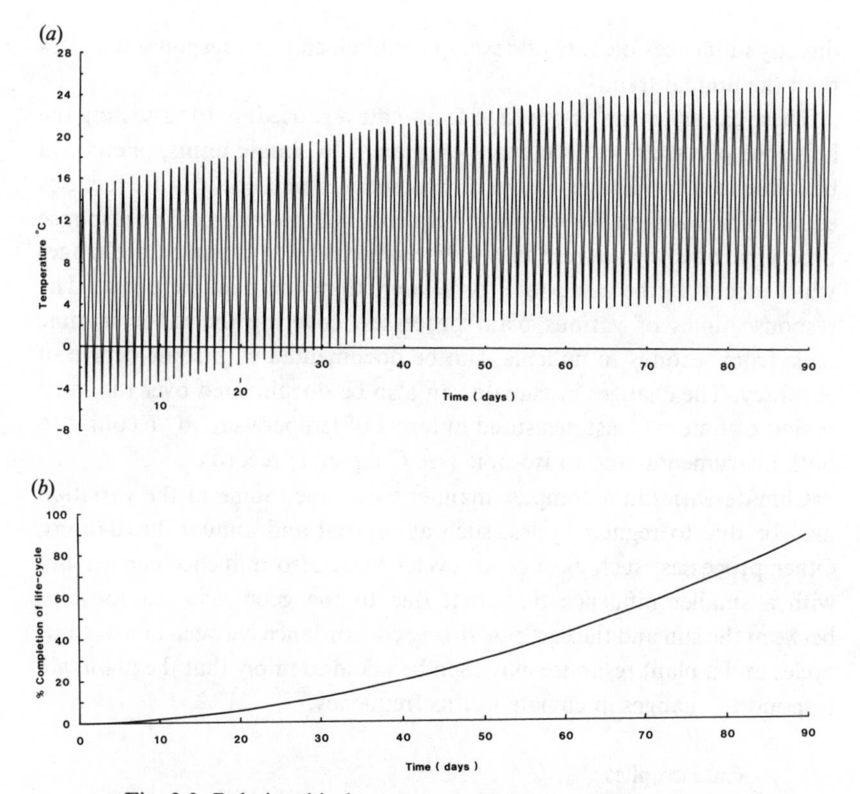

Fig. 2.3. Relationship between temperature and life-cycle completion: (*a*) temperature trend; (*b*) life-cycle completion.

The concept of the response time in these examples demonstrates that if the response time is less than the environmental response time (the reciprocal of its frequency of change) then the course of the environmental change will be followed closely. The changing environment will therefore drive the biological response.

Of course not all plant responses occur at the same frequency as the diurnal fluctuations in temperature. A typical 'annual' life cycle may be completed in about 95 days, which provides an estimate of 31 days for the response time of the cycle. The effect of this response time on the sensitivity of life-cycle development to temperature may be seen on Fig. 2.3. The diurnal variation in air temperature is superimposed on an annual variation in temperature over a period of just over 90 d (Fig. 2.3(*a*)). These variations in temperature have then been used to predict the rate of completion of a typical life-cycle, with a chosen sensitivity of 0.13% of completion/°C/day, and a response time of 31 d (Fig. 2.3(*b*)). The higher frequency diurnal trends are completely damped, whilst the annual trend directly influences the rate of development, which has a response time less than the annual trend.

These examples may be used to formulate a procedure for assessing the influence of variations in climate on plants. In simple terms, plants will be influenced by changes in the environment which occur at frequencies equal to or less than the response frequency. Or, in terms of the response time, a particular plant response will be affected by environmental changes which occur at the same rate, or slower, than the plant response. The response times of various plant processes, covering the range in time scale from seconds to millenia, can be documented with some degree of accuracy. The changes in climate can also be documented over the same period of time, at least measured in terms of temperature, by recourse to both instrumental and to isotopic (see Chapter 1) records.

Climate varies in a complex manner with time. Some of the variation may be due to regular cycles, such as diurnal and annual fluctuations. Other processes, such as sunspot cycles, may also influence climate but with a smaller influence than that due to the geometric relationships between the sun and the earth. A strong concordance between any of these cycles and a plant response may then be taken to imply that the plant will respond to changes in climate at this frequency.

Catastrophes

Wimsatt (1980) has stated that 'evolution is opportunistic and will pick that mode of response which has the best cost-benefit ratio under the circumstances'. This key statement will be further elaborated in Chapter 4. However it has a bearing here, when interpreting the responses of plants to regular cycles of climate. An annual range in air temperature from −30 °C to +30 °C would be catastrophic for the occurrence of tropical evergreen trees, which are not only frost but also chilling sensitive. However, both temperate deciduous trees and boreal evergreen (coniferous) trees can survive this wide range of temperature, as a result of particular characteristics of their life-cycles. The relevance of these responses in northern climates has been developed along the lines of the cost-benefit theory of Wimsatt discussed in Chapter 4, at the same time implying that the narrow range of temperatures which can be tolerated by tropical evergreen species is the most aggressive or cost-effective response for tropical climates.

The general climatic tolerances of trees from the boreal, temperate and tropical zones are shown on Fig. 2.4, in relation to temperature, rainfall and wind speed. The hatched areas describe the appropriate ranges of tolerance, with scales for the abscissa. The potential for describing the

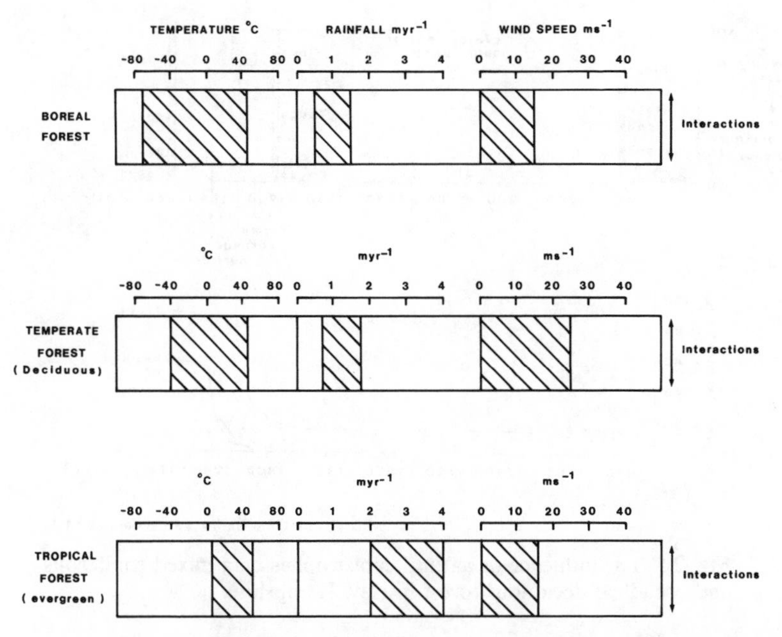

Fig. 2.4. Typical ranges of climatic tolerance of trees from different climatic zones. The hatched area represents zones of tolerance.

interaction of an additional variable is available on the ordinate. This has not been realised because of the dearth of experimental data.

The open areas on the diagram, the extremes on the abscissa, describe the ranges of climate which will be outside of the tolerances of the species, and include for example sub-zero temperatures for tropical trees, or high wind speeds for boreal conifers. When the climatic conditions fall into these ranges (as for example in frosts, fires or hurricanes) then the effects are immediate and likely to be catastrophic for the species.

Two features follow from this and the earlier analysis. First, when the climatic range is within the tolerances of the species, it can occur and compete with other species and have the capability of responding to concordant cycles of climate. Second, if the local climate exceeds the limits of tolerance, the response times are overridden and death and extinction can result. In this case, and only this case, the environmental change (i.e. catastrophe) will almost certainly have a much shorter response time than the plant process (such as longevity) which is under question.

Numerous examples of the influence of natural catastrophes have been published. However one particularly interesting set of examples is that afforded by the work of Henry & Swan (1974). This work was carried out on a small area (0.04 ha) within a forest of mixed coniferous and

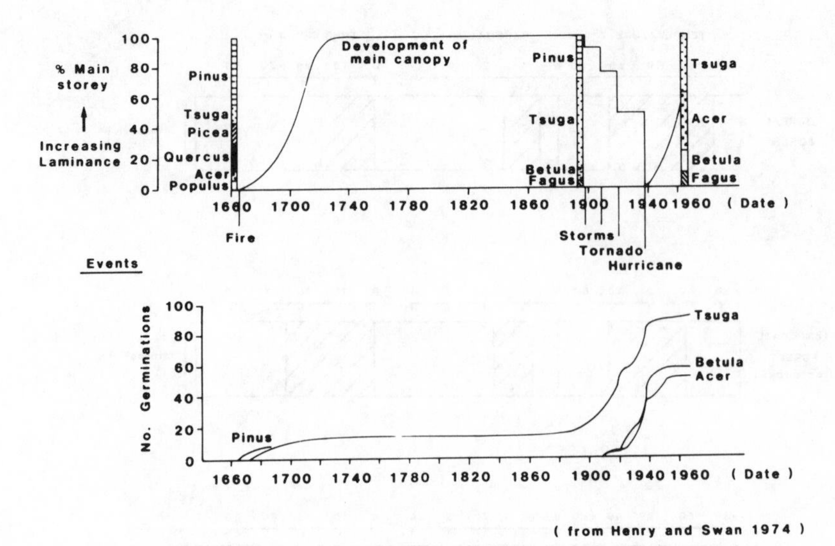

Fig. 2.5. The influence of natural catastrophes on a mixed coniferous and broadleaf deciduous forest in New Hampshire.

broad-leaved deciduous species in New Hampshire, USA. The study area had never been cut for timber and the authors were able to age both living and fallen dead trees within the study area. This information, in addition to occurrences of charcoal fragments and historical records, was used to determine the developmental history of the area over a period of 300 years from the date of a catastrophic fire in about 1665.

The times of recruitment of extant individuals from seed, and the identity, age and stature of living and fallen trees were established within the plot and have been interpreted in Fig. 2.5 to provide a developmental history of the forest.

The composition of the pre-fire forest was determined from charcoal deposits and was almost certainly dominated by *Pinus strobus*, with a range of subordinate species.

Closure of the forest canopy probably occurred some 80 years after the fire, following rapid recruitment by seed of *P. strobus*, which dominated the canopy, and *Tsuga canadensis* (Fig. 2.5). The structure and composition of the forest probably changed little up to the end of the nineteenth century, although a slow recruitment of *T. canadensis* into the understorey could be detected.

Four natural catastrophes occurred between 1898 and 1938; these were all storms increasing in violence to a tornado in 1921 and a hurricane in 1938. The hurricane effectively flattened the remaining large trees of the forest. During this period of open canopy, seed germination of *T. canadensis*, *Betula lenta*, *Acer rubrum* and *Fagus grandiflora* was abundant. The end result of this period of catastrophe was the development of a canopy in 1967 differing in character from that of the pre-catastrophe forest. The dominant species was now *T. canadensis*, with a considerable proportion of broad-leaved deciduous species, unlike any of the previous forests. *P. strobus*, originally dominant, was quite absent.

These rather infrequent (5 in 300 years) natural catastrophes, which were times when the environmental range exceeded plant tolerance, could have dramatic effects on plant distribution, in a background of tolerable climatic conditions which probably selected the local range of plant species.

In the search for the time scales of the climatic control of plant distribution, it is clear that events which occur with low probability, such as catastrophes, can exert strong influences on plant distribution. However, it is also likely that spatial scale is important in this case. In the example above, *P. strobus* became extinct in the study area after the hurricane of 1938; however, a wider view of the present forest (including both the study plot and a larger area of the native forest) shows *P. strobus* to still be an important component of the forest. The hurricane of 1938 probably failed to devastate the whole forest, particularly in a variable terrain, therefore allowing the persistence of *P. strobus* elsewhere. So it is reasonable to suggest again that the tolerable climatic conditions may select the range of species locally available, but with some local heterogeneity of composition, which may result from the random occurrences of catastrophic events.

Climatic analyses

Any lengthy period of climatic record reveals an apparently random series which may or may not be superimposed on a regular cycle of events. If climatic change drives and selects for vegetational change, then it is clearly important to distinguish between the effects of random events and repetitive cycles on plants. As established, when the random or the cyclic event exceeds the levels of plant tolerance then a catastrophe occurs. However random events may not be so extreme. How do these occurrences influence plants?

A time series of temperature, combining a regular 24 h cycle with random excursions from the cycle, is shown on Fig. 2.6(*a*). Some of the excursions are close to the same magnitude as the diurnal amplitude in temperature, whilst others are less extreme. The biological effects of this time series of temperature may be predicted after attenuation by three

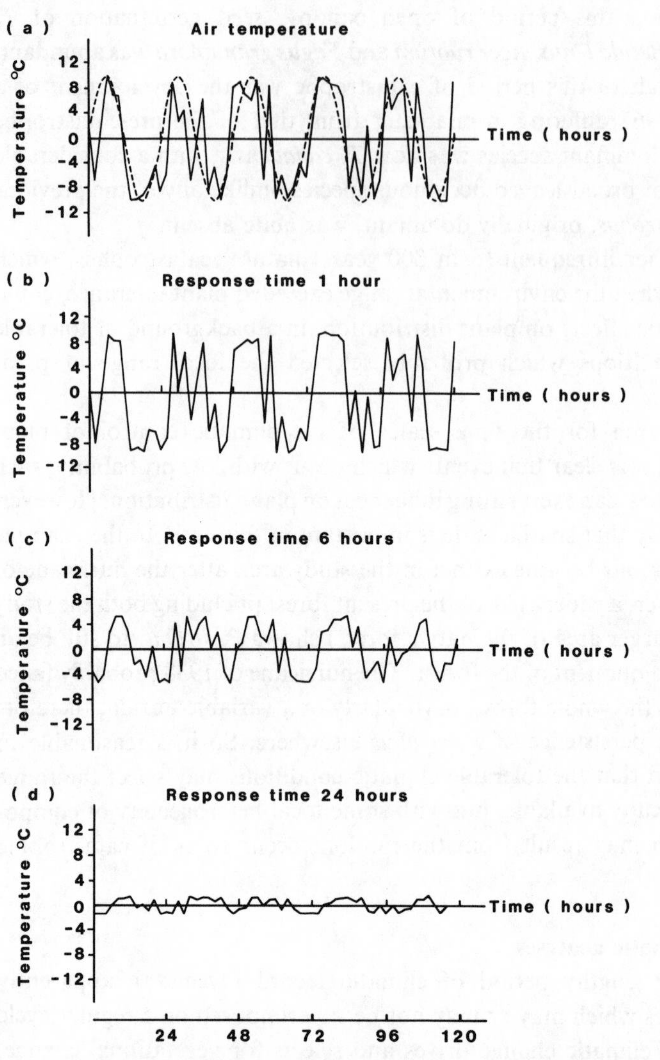

Fig. 2.6. Influence of response times on sensed temperature: (*a*) air temperature; (*b*) response time 1 hour; (*c*) response time 6 hours; (*d*) response time 24 hours.

different response times (Figs. 2.6(*b*, *c*, *d*)). A response time of 1 h has a negligible effect on the perceived temperature. However, the response time of 6 h leads to a significant attenuation of the fluctuations in temperature, effectively smoothing out the random fluctuations and revealing a more regular cycle of temperature. The response time of 24 h leads to greater smoothing and attenuation, although the cycle of temperature still dominates.

(a)

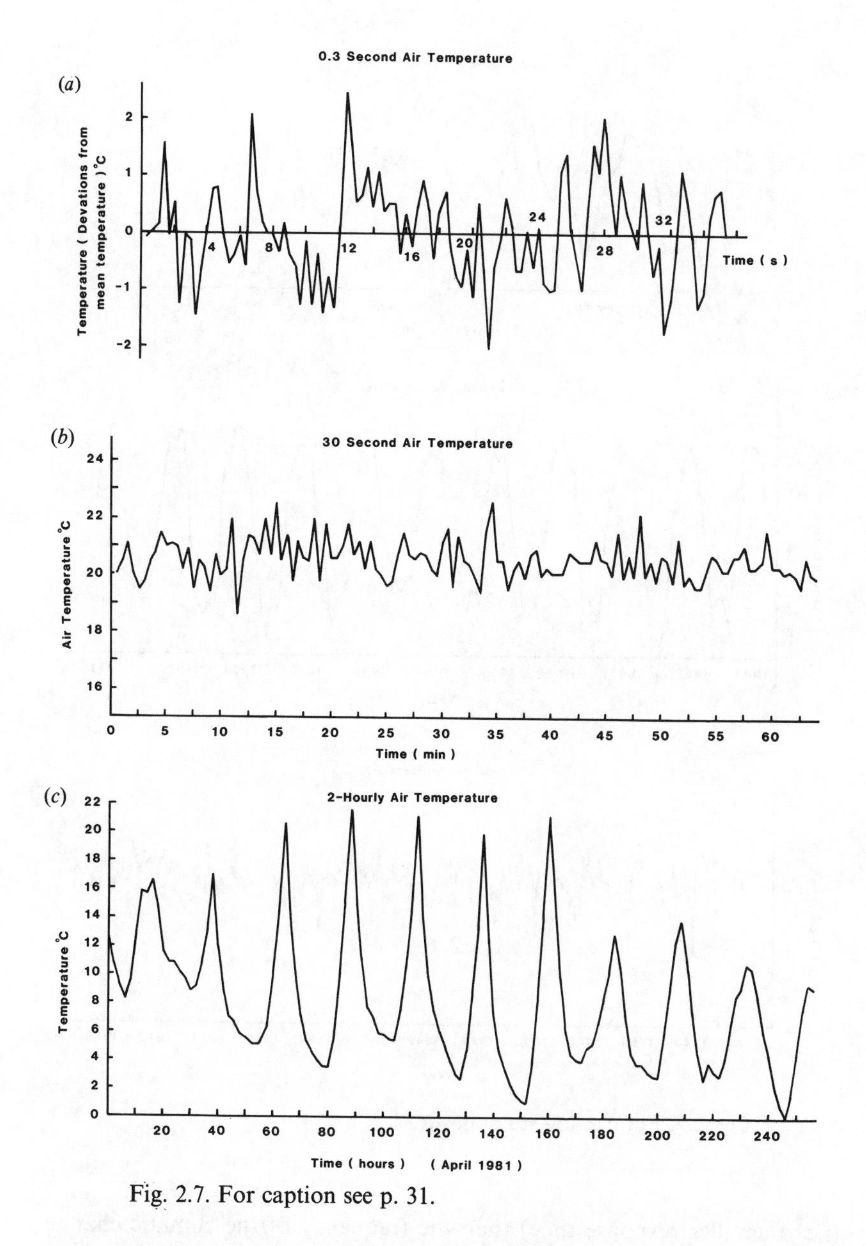

Fig. 2.7. For caption see p. 31.

The information presented on Fig. 2.6. resulting from different response times, implies that plant processes will just respond to regular cycles of climatic change when the frequency of the cycle is similar in magnitude to the response frequency (the reciprocal of response time). In fact the 'window' for this response is the steeply sloping portion of the sigmoidal response curve shown on Fig. 2.1. When the response frequency is greater

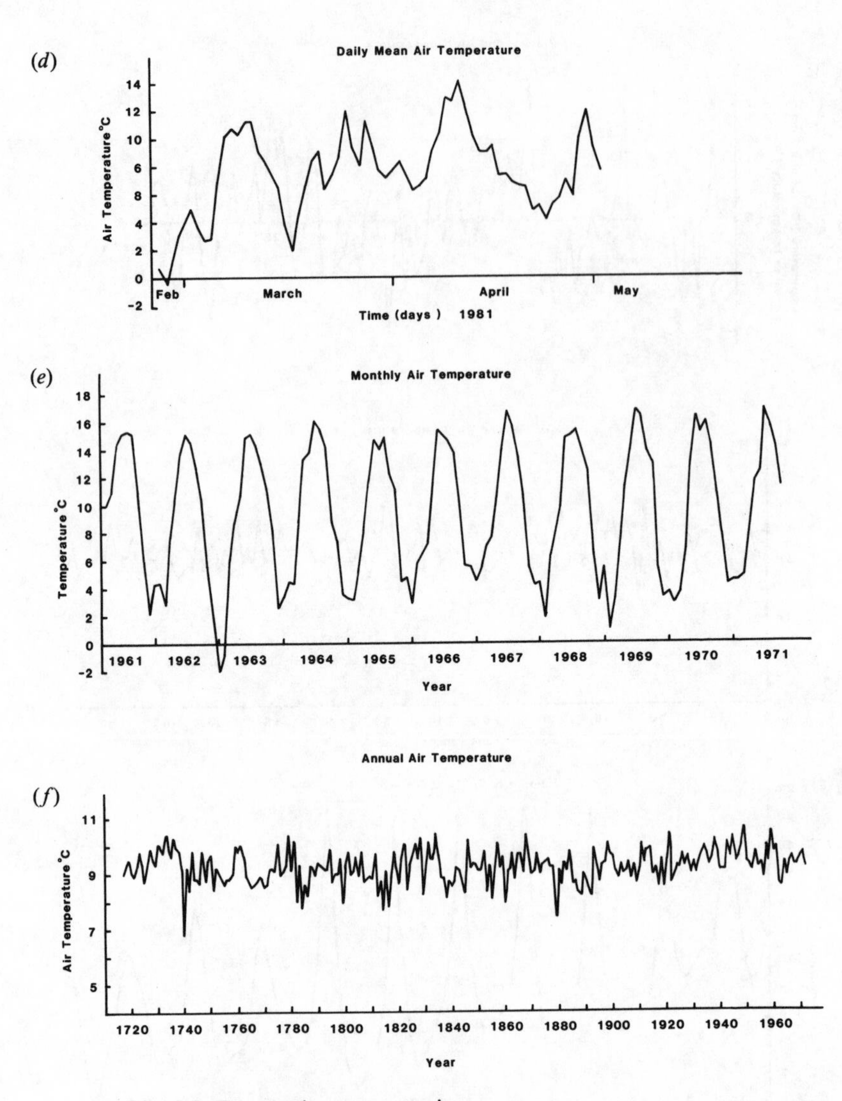

Fig. 2.7. For caption see opposite.

(i.e. a smaller response time) than the frequency of the climatic change, then the random excursions of climate will also be sensed.

With this evidence it now becomes possible to proceed with the analysis of available climatic records and to search for concordances between the cycles of climatic change and the response times of plants.

The climatic records are presented on Fig. 2.7, with a range in temporal

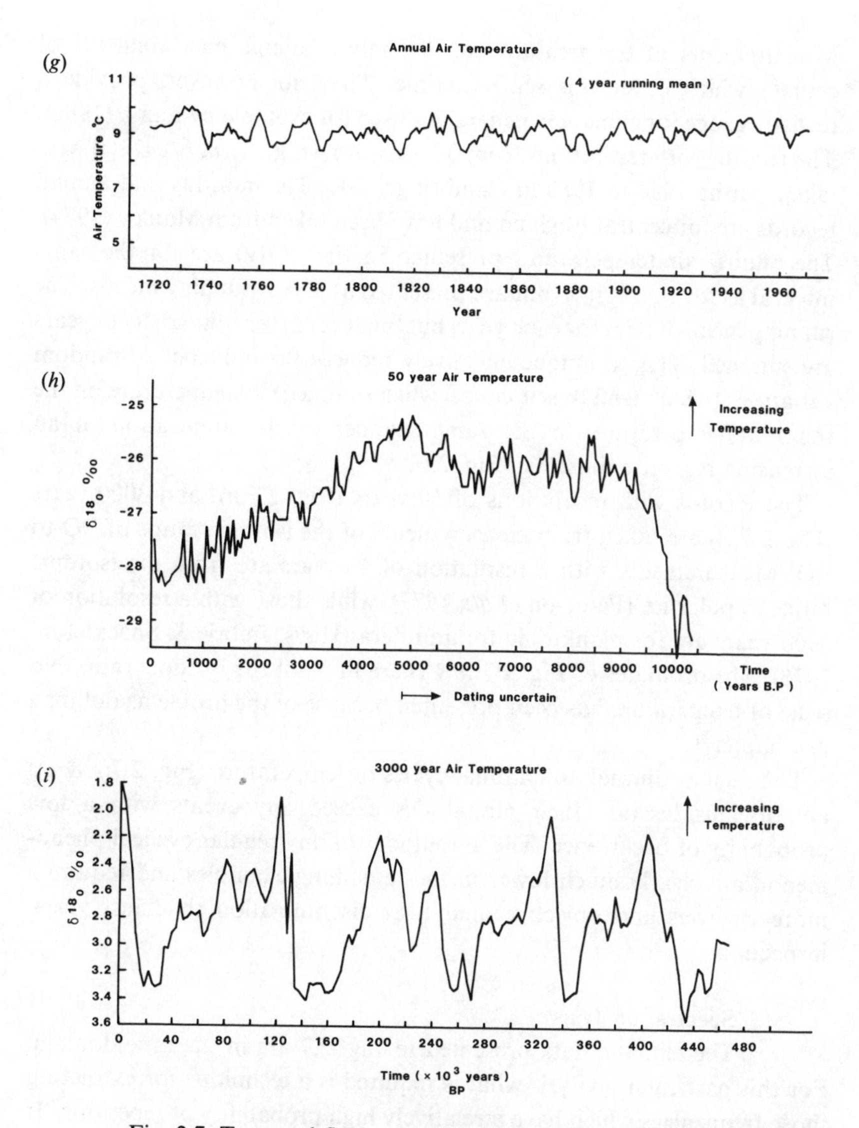

Fig. 2.7. Temporal fluctuations in temperature, with different scales of resolution: (*a*) 0.3 second resolution; (*b*) 30 second resolution; (*c*) 2 hour resolution; (*d*) 24 hour resolution; (*e*) month resolution; (*f*) year resolution; (*g*) 4 year resolution; (*h*) 50 year resolution; (*i*) 3000 year resolution.

resolution from 0.3 s to 3000 years, a range of about 12 orders of magnitude (from 10^{-1} to 10^{11} s). The last two figures (Fig. 2.7(*h*, *i*)) are from isotopic measurements with their attendant problems of interpretation (see Chapter 1), whilst the remainder are from instrumental records.

Measurements of temperature are the only available measurements of climate which cover this scale of time. They do, however, provide a realistic gauge for climate in general, as used for example by Lamb (1982). The records with resolutions from 0.3 s to 24 h (Fig. 2.7(*a*, *b*, *c* & *d*)) were taken during 1981 to 1983 in Cambridge, UK. The monthly and annual records are for central England and have been taken from Manley (1974). The annual air temperatures presented in Fig. 2.7(*g*) are for the same interval as for Fig. 2.7(*f*), but are presented as 4-year running means. The running mean applies for each year, but the records for four adjacent years are summed. This technique effectively reduces the influence of random variations (which tend to self-cancel when summed) in temperature on the trend in temperature, in the same manner as the application of an increasing response time (e.g. Fig. 2.6).

The records with resolutions of 50 years (Fig. 2.7(*h*)) and 3000 years (Fig. 2.7(*i*)) are taken from measurements of the isotopic ratios of ^{18}O to ^{16}O. Measurements with a resolution of 50 years are from the isotopic ratios in polar ice (Paterson *et al.*, 1977), while those with a resolution of 3000 years are for planktonic foraminifera (Hays, Imbrie & Shackleton, 1976). The ordinates of Fig. 2.7(*h* & *i*) are in ‰ of the isotopic ratio. No scale of temperature has been presented because of the problems outlined in Chapter 1.

The regular diurnal and annual cycles of temperature (Fig. 2.7(*c* & *e*)) are obvious because their amplitudes exceed any events with a low probability of occurrence. The amplitudes of any regular cyclical phenomenon are clearly much lower in the remaining examples and require a more rigorous and objective analytical discrimination than just visual inspection.

Spectral analysis

The climatic data presented in Fig. 2.7 are in the time domain. For this particular analysis what is required is a technique for extracting those frequencies which have a relatively high probability of repetition. If these cycles are of a sufficient amplitude then they will be discerned by those plant processes with appropriate response times (as shown, for example, in Fig. 2.6).

The appropriate method for this purpose is that of spectral Fourier analysis. The technique assumes that a time series can be constructed from a series of sinusoidal cycles, differing in frequency, amplitude and phase (the relative position on the time axis of the peaks in amplitude). The time series is analysed by extracting those cycles which account for most of the temporal changes in the time series, such as the diurnal cycle in Fig. 2.7(*c*).

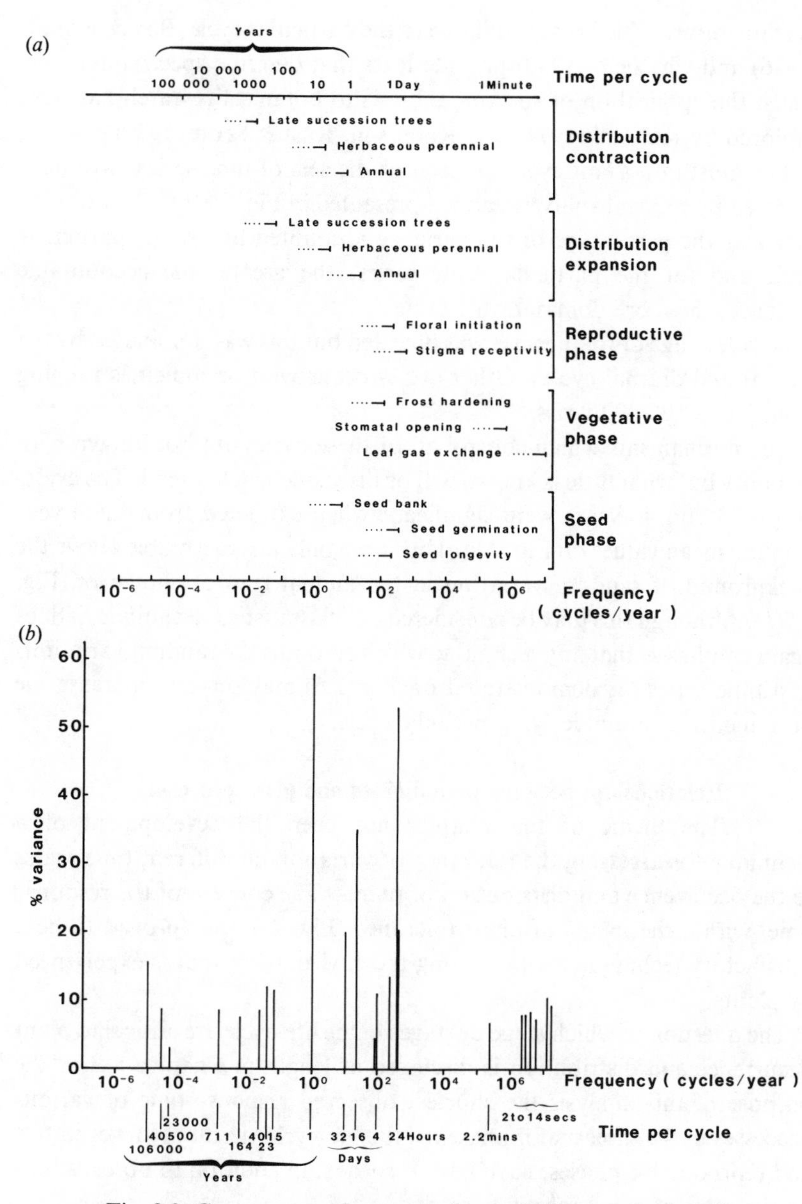

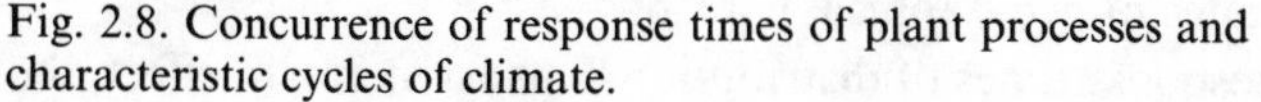

Fig. 2.8. Concurrence of response times of plant processes and characteristic cycles of climate.

In the case of the diurnal and annual cycles more than 50% of the variations in the time series are due to the one dominant cycle. In other instances more than one cycle may be important. The significance of these cycles is tested statistically in terms of that percentage of the total variance in temperature which is accountable to the particular cycle. Box & Jenkins (1976) and Chatfield (1980) provide texts that describe spectral analysis, whilst the application of spectral analysis to botanical research has been explored by Kimball (1974) and Robinson, Rust & Scott (1979).

The most important cycles in each of the sets of time series have been detected by spectral analysis and are presented in Fig. 2.8(*b*). The ordinate indicates the percentage of the variance accounted for by the particular cycle and for the particular time series; the greater the accountable variance, the more dominant the cycle.

A wide range of frequencies was detected but this was dominated by the annual and diurnal cycles. Other cycles occur with periodicities ranging from 2 s to 106 000 years.

The mechanisms which control all of these cycles are not known with certainty but what little is known will be discussed in Chapter 3. The cycles at 15, 23 and 40 years were significant when extracted from the 4-year running mean values on Fig. 2.7(*g*) but were only just detectable above the background of random variation in the annual records shown on Fig. 2.7(*f*). Although this may be considered a slick statistical technique, it does again emphasise that any technique which smooths the random variations in a time series (as demonstrated on Fig. 2.6) may serve to increase the prominence of an underlying periodic cycle.

Relationships between periodicities and plant processes

The theme of this chapter has been the development of a technique for assessing the relevance of variation on different time scales to the occurrence and distribution of plants. The concept of the response time, within the bands of plant tolerance (Fig. 2.4), has proved to be a satisfactory technique for predicting the environment that is experienced by plants.

The question of which stage or stages in the life-cycle are critical to plant occurrence and distribution is discussed at length in Chapter 5. For the purpose of this analysis the shortest observed response time of various processes in the three major phases of the life-cycle viz. the seed, vegetative and reproductive phases, have been recorded, in addition to observations on the rates of geographical spread.

The response times of distribution expansion are, in effect, the shortest times for the completion of a life-cycle. Distribution contraction takes longer and is based on the rate of species extinction.

The data have been collected from a wide range of sources, such as those referred to in Miles (1979), Li and Sakai (1982), Woodward & Sheehy (1983) and Chapter 5 in this volume with supplementary personal observations. These minimum response times are presented on Fig. 2.8(*a*), on the same time axis as the periodic cycles on Fig. 2.8(*b*). A coincidence between a dominant periodicity and a response time indicates that the cycle will be experienced by the particular process, although at a reduced amplitude (see, for example Fig. 2.6(*d*)), whilst all of the processes at shorter response times will experience the cycle and some degree of random fluctuation.

The cycles shown on Fig. 2.8(*b*) are measurements of the dominant climatic cycles, predicted from measurements of temperature. This does not imply that temperature always exerts the dominant climatic effect on plants; other correlated factors such as irradiance, rainfall or wind speed may actually be the controlling factors but which *en passant* also influence temperature.

The immediate implication of the observations presented on Fig. 2.8 is that the shortest concordance in time for geographical spread is between the annual life cycle and the 16 day climatic cycle. No cycles which are within the range of tolerance of the annual plant, and with a shorter periodicity, should exert an influence on distribution.

The times of the concordant periodicities increase with the perenniality of the plants such that, for the slow-growing, late-successional trees (Miles, 1979) only cycles with a periodicity equal to or greater than 15 years may influence geographical spread, whereas only rather longer cycles (of about 164 years or greater) may cause a contraction of the geographical range. Figs. 1.3, 1.4 and 1.5 provide evidence for the slow response of vegetation to climatic change. All of the plant processes presented on Fig. 2.8(*a*) will respond to those cycles with periodicities of 164 years and greater. In more mechanistic terms, and because plant growth is thermophilic, the expansion in the geographical ranges of species should increase during peaks of these cycles and perhaps contract in the troughs, again depending on the response time. However this simple view of plant geography will be more complex in nature, not least because of interactions between plants differing in their responses to temperature.

Plant processes with response times of about 24 h or less will respond to a wide range of climatic cycles. However the interpretation applied to Fig. 2.8 is that those responses which only contribute to a small part of the life-cycle will not influence plant distribution. This appears a strange concept when confronted with the many examples of specific responses such as seed germination, pollen tube growth or frost sensitivity (see Chapter 5), which can completely disable geographical spread. The explanation of these responses is however simple. The examples presented on Fig. 2.6 show the effect of a range of response times on discerned temperature. The dominant cycle of temperature in this case was 24 h, which was still

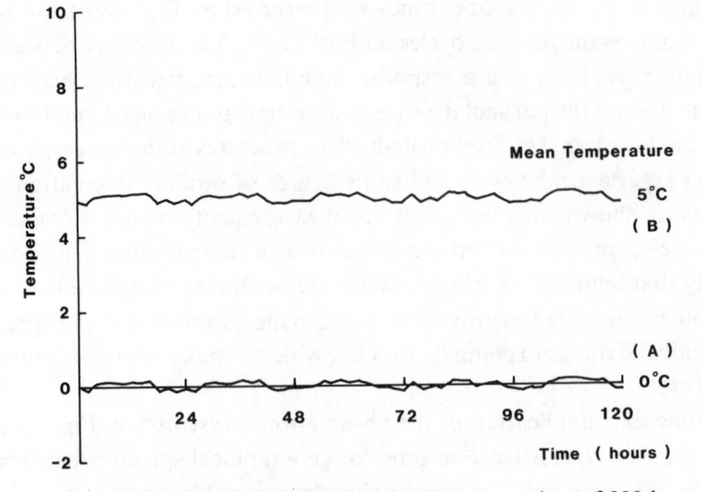

Fig. 2.9. Sensed temperature with a response time of 288 hours.

perceived at a reduced amplitude, with a response time of 24 h. When the response time is further increased to 288 h the perceived temperature is effectively constant at the environmental mean temperature of 0 °C (A on Fig. 2.9).

If the biological process which senses this temperature only occurs very slowly, or effectively not at all, at this mean temperature (such as might be the case for seed germination) then no observable response will occur. If the mean temperature increases, as a result of a much longer superimposed periodicity of climate, then over a short interval the perceived temperature might appear like B in Fig. 2.9. The higher temperature would cause an increase in the biochemical activity of the organism, perhaps stimulating the observed process such as seed germination. Empirically these effects would be interpreted as threshold phenomena. However care must be taken to separate this interpretation from just a very slow response.

Given this interpretation it becomes simple to view the blocking effect by processes with short response times on longer-term responses as that due to thresholds which in turn are sensitive to longer term cycles. This must also be considered in terms of the wide range of plant responses to an environmental change, with species appearing to differ somewhat in their response thresholds and therefore sensitivities to climatic change.

Conclusions

This chapter has been concerned with the evaluation of a technique for assessing the temporal responses of geographical distribution to climate. The concept of response times has proved useful in this respect, not only as a simple objective technique but also as an analytical tool for predicting the climatic conditions that will be experienced by plants. The major feature of this analysis is that plant processes with a particular response time will sense all cycles and random fluctuation of climate with longer response times (the reciprocal of their frequency of occurrence), but will sense negligible components of fluctuations with shorter response times (higher frequencies).

Within the normal range of tolerance of a species, geographical spread will not be influenced by processes with shorter response times than the actual process of spread, if they are operating above the apparent thresholds of response.

In this way it may be concluded that the expansion in the range of annual plants may be influenced by cycles as short as 16 days (Fig. 2.8), whilst the expansion of late successional trees may only be influenced by cycles longer than 15 years. Cyclical changes in climate are constantly occurring with a range of periodicities, implying that the geographical range, and therefore composition of vegetation, will also be constantly changing in the same manner as described in palaeoecological terms (Chapter 1), but which may not be evident from observations at the present.

References

Allen, T.F.H. (1977). Scale in microscopic algal ecology: a neglected dimension. *Phycologia*, **16**, 253–57.

Box, G.E.P. & Jenkins, G.M. (1976). *Time Series Analysis, Forecasting and Control*. San Francisco: Holden-Day.

Chatfield, C. (1980). *The Analysis of Time-series: an Introduction*. London: Chapman & Hall.

Delcourt, H.R., Delcourt, P.A. & Webb. T. III (1983). Dynamic plant ecology: the spectrum of vegetational change in space and time. *Quaternary Science Reviews*, **1**, 153–75.

Hays, J.D., Imbrie, J. & Shackleton, N.J. (1976). Variations in the Earth's orbit: pacemaker of the Ice Ages. *Science*, **194**, 1121–32.

Henry, J.D. & Swan, J.M.A. (1974). Reconstructing forest history from live and dead plant material – an approach to the study of forest succession in southwest New Hampshire. *Ecology*, **55**, 772–83.

Kimball, B.A. (1974). Smoothing data with Fourier transformations. *Agronomy Journal*, **66**, 259–62.

Lamb, H.H. (1982). *Climate History and the Modern World*. London: Methuen.

Levins, R. (1968). *Evolution in Changing Environments*. Princeton: Princeton University Press.

Levins, R. (1970). Complex systems. In *Towards a Theoretical Biology*, vol. 3, ed. C.H. Waddington, pp. 73–88. Edinburgh: University of Edinburgh Press.

Lewontin, C. (1966). Is nature probable or capricious? *Bioscience*, **16**, 25–7.

Li, P.H. & Sakai, A. (1982) (eds). Plant cold hardiness and freezing stress. In *Mechanisms and Crop Implications*, vol. 2, ed. P.H. Li & A. Sakai. New York: Academic Press.

Manley, G. (1974). Central England temperatures: monthly means 1659 to 1973. *Quarterly Journal of the Royal Meteorological Society*, **100**, 389–405.

Miles, J. (1979). Vegetation dynamics. *Outline Studies in Ecology*. London: Chapman & Hall.

Paterson, W.S.B., Koerner, R.M., Fisher, D., Johnson, S.J., Clausen, H.B., Dansgaard, W., Bucher, P. & Oeschger, H. (1977). An oxygen-isotope climatic record from the Devon Island ice cap, arctic Canada. *Nature*, **266**, 508–11.

Robinson, G.R., Rust, T.S.O. & Scott, B.I.H. (1979). Analytical approach to the study of circadian leaf oscillations in clover. I. Recording and spectral analyses of leaf oscillations. *Australian Journal of Plant Physiology*, **6**, 655–72.

Wimsatt, W.C. (1980). Randomness and perceived-randomness in evolutionary biology. *Synthèse*, **43**, 287–329.

Woodward, F.I. & Sheehy, J.E. (1983). *Principles and Measurements in Environmental Biology*. London: Butterworths.

3

World climate

Turn, fortune, turn thy wheel, and lower the proud;
turn thy wild wheel through sunshine, storm, and cloud.
A. Tennyson.

Introduction

The information presented in the previous chapters provides evidence for the considerable impact of climate on plant distribution. The records of climate appear to follow a random pattern with time, yet it is also possible to discern regular patterns within. Some of these cycles are very marked and have significant effects on plant distribution, whilst others are less marked and have less certain effects on plant distribution. Nevertheless, it is of general interest to be aware of the nature and range of climatic cycles; these will be reviewed briefly in this chapter, starting from considerations of the sun, the primary source of energy for global climate.

Solar radiation

The sun is the major source of energy for the growth of autotrophic and heterotrophic organisms. The average annual receipt of solar radiation on the earth's surface varies in more or less latitudinal bands, decreasing in the poleward directions from the equator (Budyko, 1974). This latitudinal variation in irradiance is also strongly correlated with patterns of temperature and the distribution of vegetation (eg. Cox & Moore, 1980), so it is important to investigate its control.

The relationship with latitude indicates a geometric relationship between the sun and earth. The sun, at a distance of 1.5×10^8 km from the earth, has an average temperature of about 6000K and emits electromagnetic energy with a radiant emittance of 73×10^6 Wm^{-2}. At the outer limits of the earth's atmosphere this flux density is reduced to a mean value of 1353 Wm^{-2} (the so-called solar constant, which is the irradiance on an area at right angles to the solar beam and outside the earth's atmosphere). The distance between the earth and sun varies on an annual basis by $\pm$ 1.7%, leading to variations in the solar constant of $\pm$ 45 Wm^{-2}, with a maximum at the beginning of January and a minimum at the beginning of July (NASA, 1971).

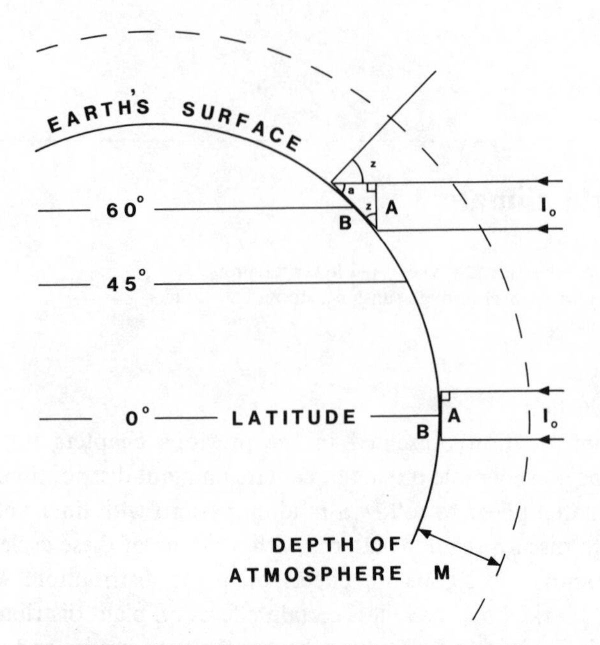

Fig. 3.1. Relationship between latitude and irradiance.

The relationship between the latitude of a site and irradiance is shown on Fig. 3.1. I_0 is the irradiance of the solar beam on a surface, B. The irradiance of surface A normal to the beam is I_0A and is the same for the two latitudes of 0° and 60°. In this instance the observations are at solar noon and at the equinox (either 21 March or 23 September), when the angle of declination, δ, the angle between the solar beam and the equatorial plane, is zero.

At the equator $B = A$, and so the irradiance of B, IB, is equal to I_0B. At a latitude of 60°, the irradiance of surface B may be described by either

$$I_B = I_0 \cos z, \tag{1}$$

where z is the solar zenith (Lambert's Cosine Law) or,

$$I_B = I_0 \sin a, \tag{2}$$

where a is the solar altitude. Under the special conditions defined for Fig. 3.1, the solar zenith is equal to the latitude ϕ, so that $z = 60°$ at a latitude of 60°.

The zenith angle of the direct solar beam may be defined under any conditions of varying latitude, declination and solar time (Gates, 1980) as:

$$\cos z = \sin\phi \sin\delta + \cos\phi \cos\delta \cos h, \tag{3}$$

where h is the hour angle of the sun, the measure of time from solar noon, where one hour equals 15°.

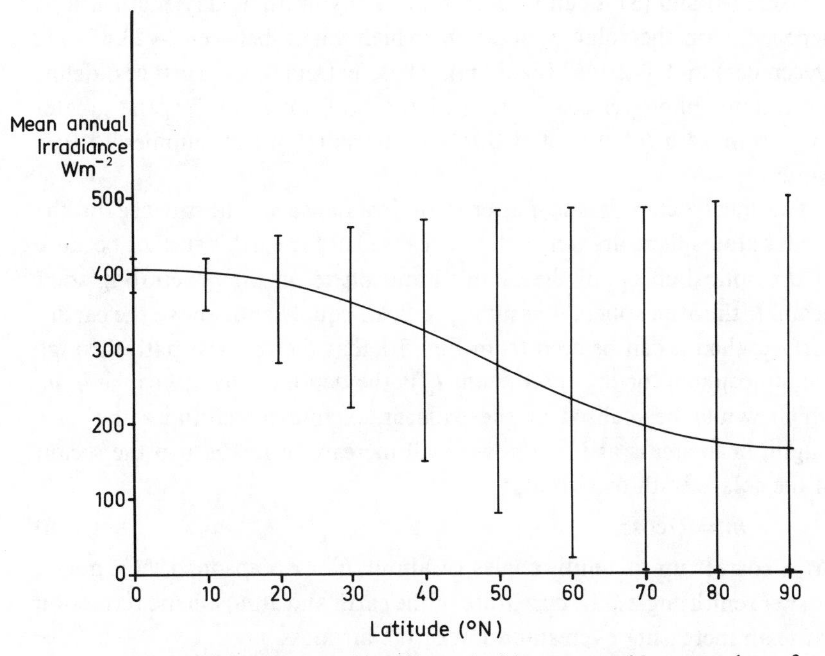

Fig. 3.2. Mean annual irradiance incident on earth's atmosphere, for the northern hemisphere, indicating maximum monthly range.

When (**3**) and the annual variation in the solar constant are taken into account, then it becomes possible to investigate the latitudinal variations of irradiance incident on the earth's atmosphere. Figure 3.2 shows the mean annual irradiance over a range of latitudes (the continuous line), with the range between the monthly extremes shown as the bars. It is clear that there are rather small latitudinal differences in mean irradiance, but considerably larger monthly ranges, which increase with latitude. The variations shown on Fig. 3.2 explain clearly the broadly latitudinal variation that can be observed beneath the atmosphere at the earth's surface.

The maximum monthly range in irradiance is also related to variations in photoperiod. The hour angle of the sun at both sunrise and sunset may be calculated from the solar declination and site latitude as:

$$\cos h = -\tan\phi \tan\delta \,, \qquad \textbf{(4)}$$

so that the time in hours between sunrise and sunset, the photoperiod, is P:

$$P = \frac{2\cos^{-1}h}{15} \qquad \textbf{(5)}$$

where $\cos^{-1}$ is the secant of the hour angle.

Using (**4**) and (**5**) it can be seen that at any latitude, daylength will be dependent on the solar declination, which varies between −23.45° (22 December) and +23.45° (22 June). These extremes, the solstices, define the extreme photoperiods, which at latitudes from about 70° and greater vary from 24 h (of potential direct solar radiation) in summer to 0 in winter.

The ideal deterministic patterns of irradiance at the surface of the earth's atmosphere are not perfectly realised at the earth's surface because of the optical effects of the earth's atmosphere on the penetrating solar beam. If the atmosphere is assumed to be of equal depth above the earth's surface, then it can be seen from Fig. 3.1 that the shortest path through the atmosphere for the solar beam I_0 is the depth of the atmosphere, m, which would be realised at the equator. At greater latitudes the path length, or air mass as it is known, will increase in relation to the secant of the solar zenith or altitude:

$$m = 1/\cos z. \qquad \textbf{(6)}$$

(**6**) is correct up to zenith angles of about 70° (Robinson, 1966), but at greater zenith angles, the curvature of the earth and atmospheric refraction cause an increasing overestimation of the air mass, m.

When the solar beam traverses the atmosphere it may be absorbed, reflected and scattered by various gaseous and aerosol components. If these optical properties are assumed to be constant with depth in the atmosphere, then the depletion of the solar beam, I_0, will be a simple function of the air mass, m, and the mean atmospheric transmittance, τ:

$$I = I_0 \tau^m \qquad \textbf{(7)}$$

In this case the attenuation of the solar beam is due to absorption, reflection and scattering. The atmospheric transmittance, τ, will depend strongly on dust and pollutants in the atmosphere and may be as low as 0.4 in a polluted atmosphere, but as high as 0.8 in very clear and dry skies (Gates, 1980).

Variations in the optical air mass may reduce the irradiance at the earth's surface, but will have a rather small influence on the latitudinal variation in irradiance. The presence of clouds and their site-related frequencies of occurrence, in direct contrast, may have considerable effects on irradiance. Monteith (1973), for example, shows that the transmission of solar radiation through nimbostratus cloud may be less than 10%, whilst the transmission of cirrus cloud may be as great as 60%. Not only the occurrence but also the type, or depth, of cloud is critical in determining the irradiance at ground level. A number of workers have derived empirical relationships which relate the irradiance at ground level

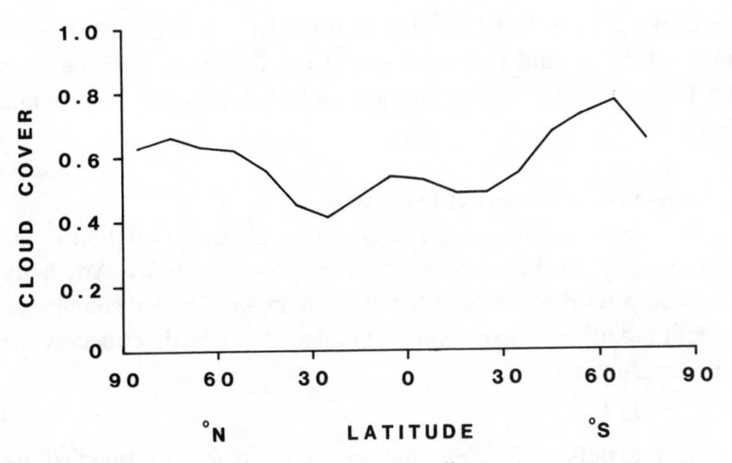

Fig. 3.3. Annual variations in cloudiness.

to a measure of cloud cover. No one relationship is perfect, at least in part because of variations in optical properties of different cloud types. However the general relationship derived by Black (1956), between the extraterrestrial solar flux, I_{ex}, and the irradiance, I, at the earth's surface, demonstrates the type of response:

$$I = I_{ex}(0.803 - 0.340f - 0.458f^2) \quad (8)$$

where f is the monthly fractional cloud cover. The relationship was derived from global observations at 88 well-separated meteorological stations. The relationship will change with variations in atmospheric transmittance, which is not included in the relationship. Nevertheless the relationship shows a marked and non-linear decrease in irradiance with cloud cover.

Variation in cloudiness will therefore be crucial in determining the global distribution of irradiance. Cloud cover will be dependent on the supply of water vapour. At one extreme, the maritime aspect of such mountain ranges as the Andes and the Rocky Mountains may have cloud cover for 80 or 90% of the year (Lamb, 1972). At the other extreme, over deserts, cloud cover may be less than one third of this (Bryson & Hare, 1974).

Latitudinal averages of cloud cover are shown on Fig. 3.3. (from Brooks, 1930). The minima in cloudiness conform with the major desert zones of the world. The maxima occur at latitudes between about 60° and 80°. The subsequent decline in cloudiness with latitude results from reduced evaporation in these very cold climates, with reduced advection from lower latitudes (Lamb, 1972).

The evidence from Fig. 3.3 suggests that the latitudinal distribution of irradiance will be modified to some extent by cloudiness, both between latitudes (Fig. 3.3) and within latitudes, such as seen for deserts and orographic zones.

Long-wave or terrestrial radiation

One crucial feature of the interception of solar radiation is the subsequent increase in the temperature of the absorbing body. Any body which absorbs radiation and has a temperature greater than absolute zero will also emit radiation. This emission of radiation can be described by the Stefan–Boltzmann law:

$$R = \varepsilon\sigma T^4, \tag{9}$$

where R is the radiant flux emitted per unit area, σ is the Stefan–Boltzmann constant (5.6697×10^{-8} Wm^{-2} K^{-4}) and T is the absolute temperature measured in kelvins. The emissivity, ε, is a measure of the efficiency of emission, which for most natural objects such as water, soil or vegetation is of a high value at about 0.9 (or greater). In addition, a good emitter of radiation is also a good absorber of radiation (Kirchoff's Law). Assuming an emissivity of unity, according to (**9**), the radiant emission R for the sun, with a temperature of 6000K, is 73.5×10^6 Wm^{-2}, whilst R for a terrestrial body at 293K is 418 Wm^{-2}. The very high solar radiance is reduced considerably by the time it reaches the earth's atmosphere and, at this point, the irradiance is the solar constant, which at 1353 Wm^{-2}, is of a similar magnitude to the radiation emitted by the terrestrial body. It is clear therefore, that both solar and terrestrial radiation will be critical in determining the thermal environment of the world.

Radiation emitted by terrestrial bodies differs from solar radiation in terms of the wavelengths of emission. The wavelength of peak emission, λ, may be defined by Wien's Law:

$$\lambda = 2897/T \tag{10}$$

where λ is in μm and T in kelvin. At 6000K, λ has a value of 0.48 μm, which increases to 9.89 μm at 293K (20 °C).

The waveband of emission is also broadly related to the wavelength of peak emission. For short-wave solar radiation the waveband extends between about 0.3 μm and 4 μm, but for terrestrial or long-wave radiation the waveband extends between about 3 μm and 100 μm. The irradiance per unit wavelength will be significantly larger for solar radiation than terrestrial radiation, a feature which is critical for controlling photo-

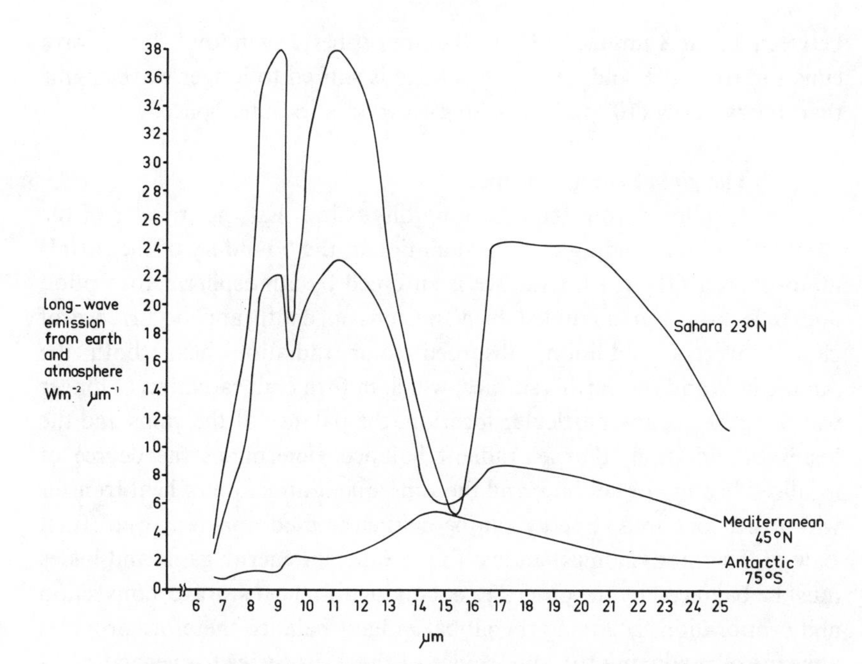

Fig. 3.4. Upward flux of long-wave radiation from the earth and clear atmosphere, measured from a satellite (from Hanel *et al.*, 1971).

chemical processes which require high energy levels (such as photosynthesis, photomorphogenesis and vision). In general, the impact of terrestrial radiation is on the thermal or energy balance of organisms, where wavelength-independent absorption of radiation is the rule.

The relationships between long-wave emission, surface temperature and latitude may be seen on Fig. 3.4. The diagram shows the long-wave radiation emitted from the earth's surface and from the atmosphere to a satellite receiver outside the atmosphere. The diagram is a simplification of the results presented by Hanel *et al.* (1971) and shows the long-wave emission from three latitudes, 23° N, 45° N and 75° S. The emission temperature for these curves is not known and, in addition, such a measure is complicated by the fact that it is a function of the temperature of both the ground surface and of the atmosphere. However, in relative terms, it is clear that temperature and long-wave emission decrease markedly with latitude.

The emission spectra for the Sahara and the Mediterranean also show peaks and troughs of irradiance, due to absorption by the gaseous components of the atmosphere (in particular water vapour, carbon dioxide, oxygen and ozone). Minimal absorption occurs in the waveband between about 8 μm and 13 μm, the atmospheric 'window'. Long-wave emission from the cold Antarctic surface is shifted to longer wavelengths than the window (**10**), reducing long-wave loss to outer space.

The global energy balance

It follows from the preceding discussion that the transfer of the strict latitudinal banding of solar radiation at the boundary of the earth's atmosphere to the earth's surface is enforced by atmospheric absorption and reflection, but disrupted by variations in depth and occurrence of cloud cover. In addition, absorbed solar radiation heats both the atmosphere and the earth's surface, which in turn emit radiation at longer wavelengths. At any particular location, the balance of the gains and the losses of radiation, the net radiant balance, determines the degree of localised heating or cooling, and the subsequent direction of heat transfer with other locations. Energy can be neither created nor destroyed (First Law of Thermodynamics) and so this balance of energy gains and losses must be balanced by the processes of conduction, heat storage, convection and evaporation. Mapping the global radiant balance therefore provides a means of predicting the importance of these processes for vegetation at different geographical locations.

The average global balance is shown in Fig. 3.5, and is based on information presented in Lamb (1972), McIntosh & Thom (1978), Oke (1978) & Sorensen (1979). Figure 3.5(*a*) shows the upward and downward fluxes of both short-wave and long-wave radiation, as a percentage of the mean extraterrestrial flux, which for the whole globe has an average of about 350 Wm^{-2}. Figure 3.5(*b*) shows the fluxes in units of Wm^{-2}.

The left side of each diagram shows the fate of the short-wave or solar radiation; the right side shows the long-wave or terrestrial radiation. The reflection of the incoming solar extraterrestrial flux to outer space is 27%, of which 19% is by reflection from clouds (Sr_c), 5% from the atmosphere (Sr) and 3% from the earth's surface (Sr_s). A more or less equal proportion (26%) is absorbed (Sa) by the atmosphere (20%) and by clouds (6%). The remaining 47% is absorbed at the earth's surface.

In comparison, the fluxes of long-wave radiation are a little more complex. The average emitted flux from the earth's surface (Lu_s) is 111% of the extraterrestrial flux and on average only 4% of this passes through the atmospheric window to outer space (Lu_w). Of the remainder, 79% is absorbed (La) by the atmosphere, and 28% by clouds. A greater proportion of the long-wave emission from the atmosphere and clouds is towards the earth's surface than towards outer space. This is because of the greater atmospheric density and temperature closer to the earth's

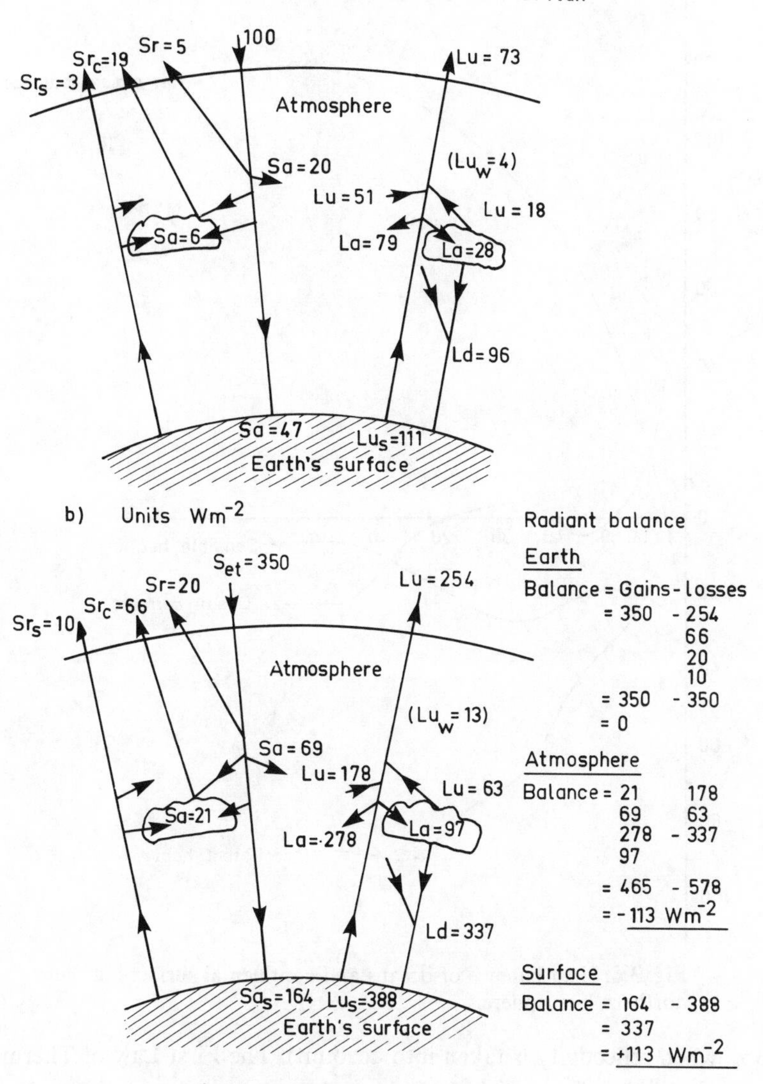

Fig. 3.5. Average energy balance of the earth.

surface. As a result the downward flux of long-wave radiation from the atmosphere and clouds (Ld) is 96% whilst the total upward flux to outer space is 69% (Lu).

A total equal to 73% of the extraterrestrial flux of solar radiation is emitted as long-wave radiation by the earth's surface and atmosphere to outer space. This exactly balances the incoming solar radiation (which is

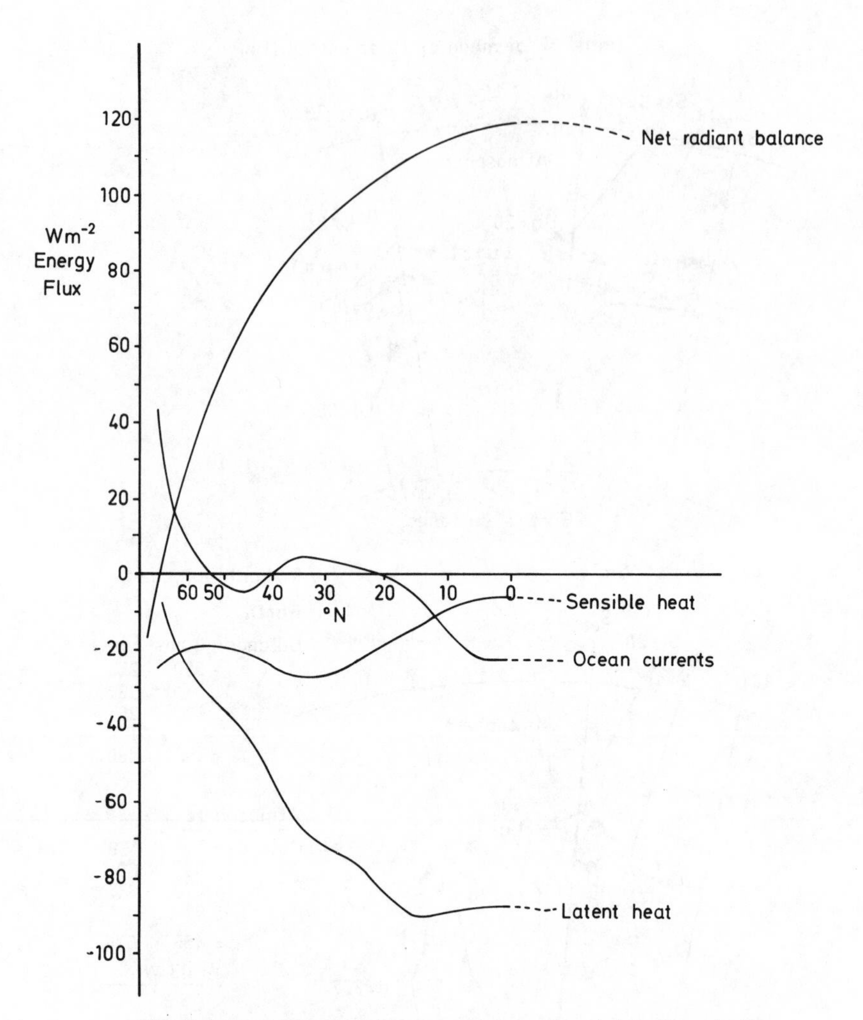

Fig. 3.6. Heat balance of oceans and continental surfaces in the northern hemisphere.

73% when reflectivity is taken into account). The First Law of Thermodynamics is therefore satisfied.

The lower diagram and balance sheet demonstrate the differences in the net radiant balance of the earth plus atmosphere, which is exactly zero, the earth's surface alone, which has a positive balance of 113 Wm^{-2}, and the atmosphere, which has a negative balance of -113 Wm^{-2}.

The radiant balances which differ from zero must be balanced in terms of the processes of evaporation, convection, conduction and heat storage. A latitudinal view of these processes is shown in Figs. 3.6 – 3.8, using data from Budyko (1955) presented in Lamb (1972). All three figures are for

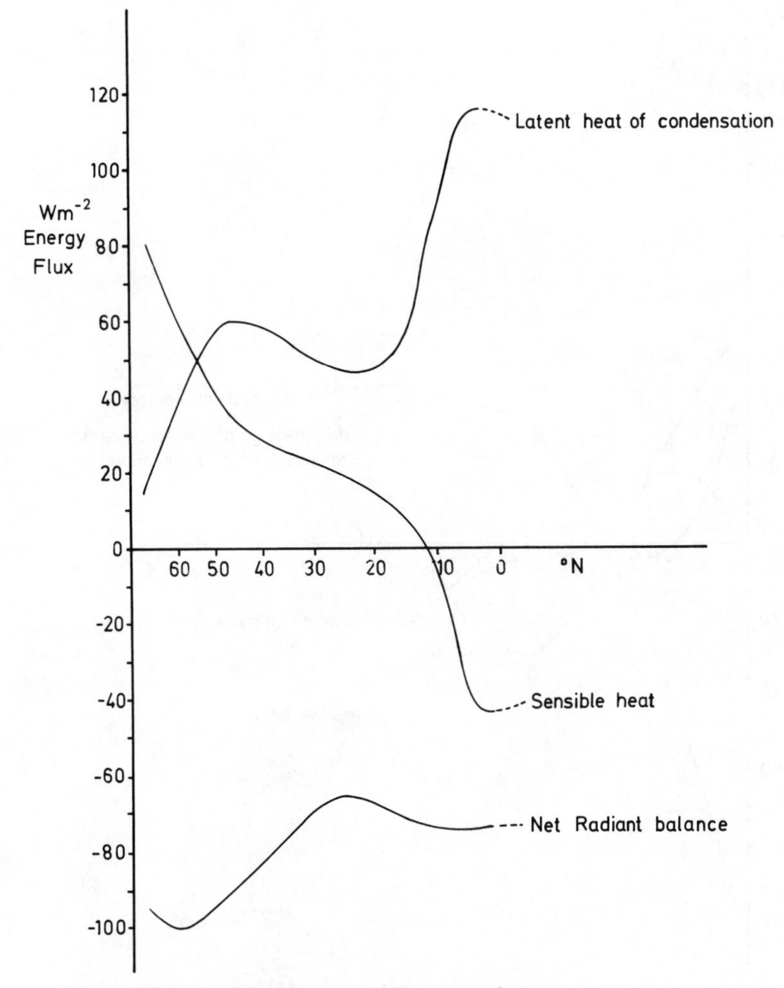

Fig. 3.7. Heat balance of the atmosphere.

the northern hemisphere, with fluxes of energy plotted against the sine of the latitude. Unit integrated area on the figures therefore corresponds to a flux of energy per unit area, at any latitude.

The net radiant balance of the oceans and continental surfaces is shown in Fig. 3.6. The radiant balance declines with latitude, falling to zero at about 80 ° N. This net radiant balance, R_N, is balanced in an energy budget by latent heat transfer E (evapotranspiration), sensible heat transfer C (convection), and heat storage in the oceans, followed by the transfer of this heat to other latitudes by ocean currents (G):

$$R_N = E + C + G \tag{11}$$

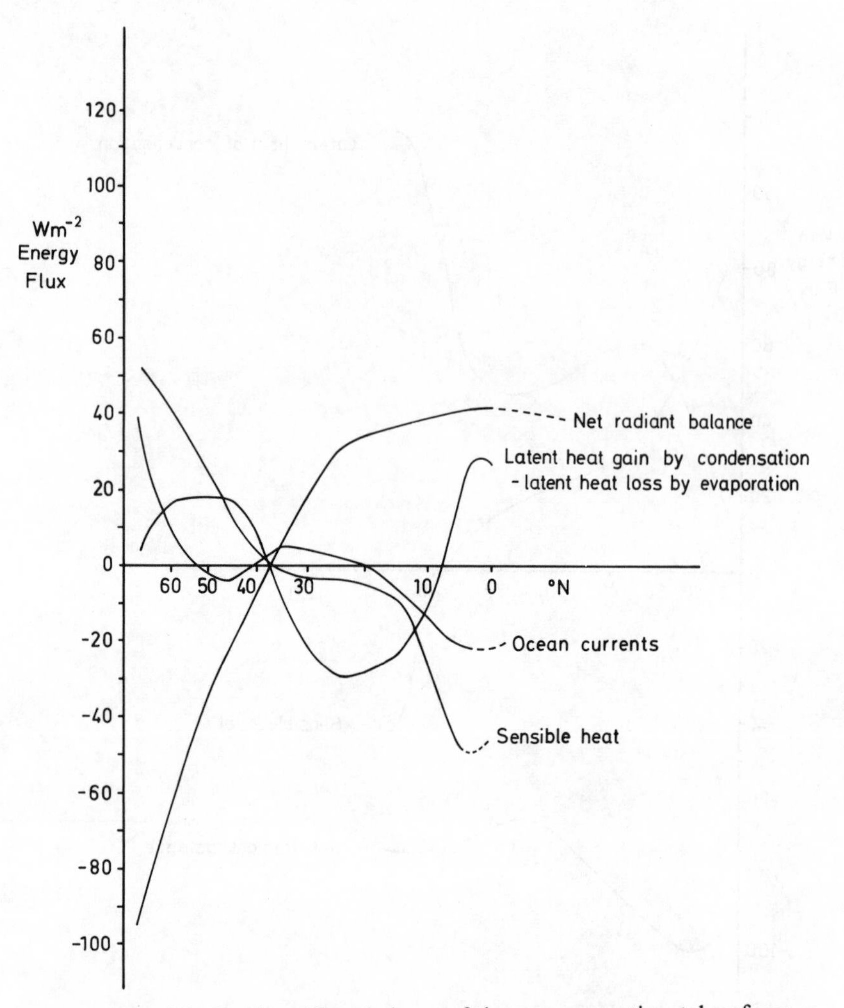

Fig. 3.8. Combined heat balance of the oceans, continental surfaces and the atmosphere.

The majority of a positive net radiant balance is dissipated by evaporation or latent heat transfer. The evaporation of water is energy requiring, extracting 2450 $J\ g^{-1}$ at 20 °C. Evaporation therefore has a cooling effect, in direct opposition to the heating effect of a positive radiant balance. Latent heat transfer is therefore negative in sign on Fig. 3.6.

Sensible heat loss is negative at all regions and reflects the convective loss of heat from the warm oceans and continental surfaces to the colder atmosphere above.

The importance of the influx of stored heat by ocean currents is clear to see at the high latitudes. At these latitudes, where the radiant balance is just positive or even negative, this represents the major source of heat (positive in sign) for maintaining the temperature at a higher level than would otherwise be the case.

The net radiant balance of the atmosphere is negative at all latitudes (Fig. 3.7). In the equatorial zone, which includes the regions of wet tropical forests, there is abundant latent heat transfer from the earth's surface to the atmosphere. The potential energy is only realised as heat for warming the atmosphere when the water vapour condenses, releasing the latent heat of condensation. Very large quantities of heat are released in this way in the equatorial region, but this drops markedly between latitudes 10° N and 30° N, coinciding with the latitudes of the major deserts of the world (where little evaporation occurs). At latitudes greater than 10° N, sensible heat transfer becomes an increasingly important source of energy for balancing the energy budget. Sensible heat is transferred vertically from the earth's surfaces beneath (Fig. 3.6) and also by horizontal transfer, or advection, in the major circulation patterns of the earth. These patterns of wind movement are not only critical in maintaining temperature at high latitudes, but they are also important in driving ocean currents, by absorption of the wind's momentum.

The final diagram of energy exchange (Fig. 3.8) combines the energy balances of the ocean and continental surfaces with the atmosphere. A regular, latitudinal change in net radiant balance is clearly in evidence, declining to zero at about 37° N, and becoming increasingly negative at higher latitudes. There is no evidence for regular latitudinal patterns in other features of the energy budget. For instance, the net energy released from evaporation (latent heat gain by condensation less the latent heat loss by evaporation), is positive in the equatorial regions and at latitudes greater than about 40° N, but negative in between. In the equatorial region therefore, there are two sources of heat, the net radiation load and the net latent heat, which are in turn balanced by significant transfer of energy towards higher latitudes by sensible heat and ocean currents. These latter features are clearly decisive in maintaining a supply of energy for high latitudes with a negative radiant balance.

Climatic change

The observations on Fig. 3.8 show that even when arranged on a latitudinal basis, there are distinctly complex patterns of energy transfer between latitudes. It is also easy to see that if there are changes in the climate of one large area, or the globe as a whole, then considerable latitudinal changes in the energy budget may result. Observations and explanations for these changes, and changes in climate, have been of considerable interest, particularly since the start of instrumental records of climate. A brief analysis of climatic records, measured as temperature or temperature related effects and spanning 13 magnitudes of scale in time, has been described in Chapter 2. It was clear from this analysis that some variations of climate are repetitive, alongside evidently random occurrences. Repetitive events or cycles may best be recognised in the diurnal and annual fluctuations of climate. These cycles are deterministic, with mechanisms which are controlled by the rigid geometries of the daily rotation of the earth and the annual variation in the inclination of the earth's axis of rotation.

Other variations in climate, such as variations with a time scale in seconds or minutes, may result from the structure of eddies over the vegetation. These eddies vary in size, structure and appearance and are clearly not deterministic in the sense of an invariant geometry. The occurrence of variations in climate controlled this way can only be demonstrated by recourse to statistical techniques, and may therefore be described as probabilistic or stochastic.

Mitchell (1976) has provided a logical framework for describing variations in climate. He defines those processes which are external to the climatic system of the earth's surface and atmosphere, such as variations in the radiant emittance of the sun, as external, forcing mechanisms. The processes within the climatic system which lead to climatic change were defined as internal, stochastic mechanisms. The implication from these definitions is that extrinsic mechanisms, related to the geometry of the earth and sun, or to variations in solar activity, are deterministic. The effects of volcanic eruptions were also considered to have extrinsic mechanisms, although the timing of eruptions is clearly probabilistic (Lamb, 1972).

The changes resulting from extrinsic mechanisms may also be modified by intrinsic mechanisms, indicating a dependence between the two processes. Although the deterministic variations are generally simpler to discover, it does not follow that they may have the greatest effect on plant distribution. It follows from the discussions of threshold phenomena, in Chapter 2, that it is both the amplitude and period of change that are critical; one random cycle may be as efficient as one regular cycle in affecting plant distribution.

Deterministic mechanisms of climatic change

The most obvious deterministic mechanisms of climatic change are the diurnal and annual changes described above. Brier (1964) describes a cycle of about 14 days which has been traced to lunar tidal influences. The effect can be observed as variations in cloudiness and rainfall, but the influence on plant distribution for any species other than short-lived annuals is likely to be small (Chapter 2).

A favourite area of investigation for cycles is in the occurrence of solar sunspots. Sunspots are areas on the sun's surface of lower-than average temperature, but of intense magnetic and solar activity Kondratyev & Nikolsky (1970) have shown that the solar constant also varies by as much as 2% during sunspot cycles. The period between successive sunspot minima has a mean interval of 11 years (Schneider & Mass, 1975). A cycle of 22 or 23 years, the double sunspot cycle, can also be recognised and is associated with reversals of the solar magnetic field (Currie, 1974; Pearson, 1978). In addition to the 11 and 22 or 23 year cycles, sunspot cycles of about 45, 80, 150, 200, 500 and 1000 years have all been reported (Gleissberg, 1958; Dewey, 1960; Eddy, 1976). The impact of these cycles will depend on their effectiveness in perturbing the global energy balance. Schuurmans (1981), for example, considers that the effect of variations in the sunspot cycle may be realised as changes in atmospheric circulation.

The effect of these cycles on plant distribution is not always clear. Perhaps the best circumstantial evidence is that at the most extreme period of the Little Ice Age in the mid seventeenth to early eighteenth century, during which time species extinctions occurred in Europe (Lamb, 1982), there was a near absence of sunspots, a period known as the Maunder Minimum (Eddy, 1976).

In some, but certainly not all, cases there are clear correlations between tree growth and sunspot numbers or cycles (positive correlations: Outi, 1961; Bray, 1968; Pearson, 1978; Sonett & Suess, 1984; no correlations: Bryson & Dutton, 1961; Stuiver, 1980).

Recently, Wijmstra *et al.* (1984) have provided palaeoecological evidence for cyclical variations in pollen production by *Alnus*, *Quercus* and *Corylus*. However, it is not known if any significant changes in geographical distribution were involved. Abbot (1963) has shown changes in plant occurrence during the severe droughts in the American great plains and these changes correlate with sunspot cycles of 46 and 91 years.

Other regular extrinsic effects on world climate which have attracted attention are those concerned with the very long-term variations in the earth's orbit, probably due to gravitational interactions with the sun, moon and planets. Three major variations in geometry have been considered, and their description and climatic effects have been described by

Milankovitch (1920, 1930, 1941). He proposed, in the same way as Adhémar (1842), Croll (1875) and others before him, that such variations in geometry could lead to the major ice ages of the world.

Milankovitch proposed that three variations in the earth's orbital motion would occur, with periodicities of about 100 000, 40 000 and 21 000 years. The longest periodicity of 100 000 years entails a change in the shape of the earth's orbit of the sun, from almost circular to markedly elliptical, and back to circular. At present the orbit is nearly circular and so the difference between the perihelion (the shortest distance from the earth to the sun) and the aphelion (the longest distance) is about 3.5%. In its most eccentric (elliptical) orbit, the effect on differences in solar constant at the aphelion and perihelion may be as large as 30% (Pearson, 1978).

The second cycle, with a periodicity of about 40 000 years, is due to continuous variations in the tilt of the earth's axis of rotation, within the range of 21.8° and 24.4° (23.4° at present). An increase in the angle of tilt leads to an increase in the contrast between summer and winter climate.

The shortest cycle, of 21 000 years, relates to the observed hemispherical differences in the solar constant. At present, the earth's orbit is nearly circular, and so the solar constant changes rather little through the year. Even so, it varies from a maximum in December to a minimum in July. This is because the perihelion of the earth to the sun is in December, with the aphelion in July. The timing of the perihelion and aphelion changes because the earth's axis wobbles. The result of this wobble in 10 000 years time will be a change from the perihelion occurring during summer in the southern hemisphere, to during summer in the northern hemisphere. This will cause a shift, in 10 000 years, to a more extreme difference between summer and winter climates.

The effect of these orbital variations is not one of a variation in incoming irradiance but in its distribution, in particular between the hemispheres. Of course, the three cycles occur simultaneously but not synchronously, leading to a long and complex time series of climatic variation.

The effects of these cycles of change may only be small; however Calder (1974) has calculated that a fall in summer insolation of only 2% may be sufficient to allow the spread of ice. Should the spread of ice be induced in this way, probably at latitudes between 50° N and 70° N, then it may be possible for amplification of the insolation effect. This would occur through an increase in the areas under ice, and therefore an increased reflectivity of solar radiation away from the earth's surface, leading to a reduction in the absorption of solar radiation and a decline in temperature (Lamb, 1982). Pisias & Shackleton (1984) have analysed concurrent measurements of oxygen isotopes relating to the volume of global ice (see Chapter 1, this volume) and estimates of global carbon dioxide, for a 150 000 year core taken from the equatorial Pacific (Shackleton *et al.*, 1983). They conclude from their analyses that changes in atmospheric carbon dioxide occurred simultaneously with orbital variations and probably amplified the small changes in insolation. The carbon dioxide concentration increased in step with the decline in the ice sheets, which could lead to an amplification of the variations in insolation (Milankovitch cycles) by increasing the downward flux of long-wave radiation to the earth's surface. This might occur through an increase in the carbon dioxide concentration leading to an increase in atmospheric absorptivity to long-wave radiation. The increased absorptivity also leads to an increased long-wave emissivity to the earth's surface (Plass, 1956), with both effects increasing the net radiant load.

Considerable evidence has now accumulated in favour of the model proposed by Milankovitch. This is provided by records of variations in sea-level preserved in coral terraces (Broecker *et al.*, 1968 and Mesollela *et al.*, 1969); from careful statistical analyses of oxygen isotopes in deep-sea foraminifera (Hays, Imbrie & Shackleton, 1976, see Chapter 1); from profiles of loess (dust blown from the periglacial areas adjacent to glacial margins) in China (Lu, 1981, referenced in Goudie, 1983) and from insolation related changes in the altitudinal limits of vegetation (Davis, 1984). Imbrie & Imbrie (1979) and Goreau (1980) have emphasised strongly the importance of orbital variations in leading to significant changes in both climate and plant distribution (and e.g. Chapters 1 and 2, this volume).

In addition to observations in regular cycles of extrinsic climate control, there is also evidence for an impact on climate by volcanic eruptions. Lamb (1970, 1971), Cadle, Kiang & Lonis, (1976) and Russell *et al.* (1976) have all provided considerable evidence for a direct impact on global climate resulting from the explosive eruption of volcanic ash and sulphuric acid droplets into the stratosphere. The droplets in particular, and the ash to a lesser extent, may have residence times in the stratosphere and atmosphere in the order of a few years, with residence time increasing with decreasing particle size. In addition, stratospheric winds will spread the particles over a range of latitudes, dependent on the pattern of global circulation. Lockwood (1979) suggests that volcanic dust injected into the equatorial zone of the stratosphere will be spread over the whole of the globe, perhaps accumulating over the polar ice caps. However dust which is injected in high latitudes will fail to spread to lower latitudes than about 30 ° because of the global patterns of circulation. This apparent tendency for particles to remain above high latitudes will result in the maximum effect on world climate because the effect on incoming solar radiation increases with latitude (Budyko, 1974) due to greater reflectance with increasing angles of incidence (Fig. 3.1).

The increase in volcanic particulates in the stratosphere will cause an increase in the reflection of solar radiation away from the earth to outer space. Dyer & Hicks (1965), for example, show a 20% reduction in the irradiance of the direct solar beam at Melbourne, Australia after the eruption of Mount Agung in Bali, during 1963. However, the total income of both direct and diffuse (scattered) radiation was reduced by only about 5–10%, due to an increase in the diffuse component of radiation.

The effect on global climate of volcanic eruptions may be rather small. For example, when Krakatoa erupted in 1883, about 50×10^6 tonnes of dust were injected into the atmosphere, leading to a reduction in mean global temperature of only 0.5 °C (Ramade, 1984). Kelly & Sear (1984) have shown that the cooling effect of a cloud of volcanic ash is very rapid, occurring within 1–3 months of eruption. Lamb (1971) has shown, for the British Isles, that many of the coldest and wettest summers (e.g. 1695, 1725, 1816, 1879, 1903 and 1912) occurred at times when the volcanic dust content of the stratosphere and atmosphere was high.

The effects of these changes on plants and plant distribution are not clear. One mechanism of change has been considered by LaMarche & Hirschboeck (1984), who propose that zones of frost damage may occur in the annual rings of species such as *Pinus longaeva* and *P. aristata* within 1 or 2 years of the occurrence of volcanic eruptions. This suggests that the volcanic effects may influence plant distribution through a lethal threshold phenomenon. The importance of minimum temperatures in the control of plant distribution is also proposed in Chapter 4.

Bray (1974) has proposed that volcanic activity may also have been the trigger for relatively recent glacial advances, such as in the periods of 5400–4700 BP, 2850–2150 BP and 470–50 BP, with a direct effect on plant distribution. It is also possible that the injection of volcanic ash into the atmosphere during a sensitive stage of the Milankovitch cycles may be an efficient technique for hastening the approach to an ice age.

Stochastic mechanisms of climatic change

The discussion of deterministic or extrinsic mechanisms of climatic change has shown the often extreme difficulty in achieving an unequivocal recognition of the deterministic control of climate. This arises in part because the extrinsic event is in some way modified by intrinsic variation, (or earth to atmosphere interactions). A single stochastic event (with a long or short time scale) may be as important for the control of plant distribution as a deterministic event, so it is important to recognise the manner in which these events might occur.

Perhaps the most familiar stochastic events are the changes in synoptic weather patterns, typified by the movement of regions of high and low atmospheric pressure. The occurrence of such areas of pressure is limited to a few days at the most and are only likely to exert an effect on plant distribution if they lead to extreme climatic conditions.

The spatial and temporal occurrences of these high-and low-pressure systems are controlled by the great flows of the circumpolar vortex. In each hemisphere of the globe, and through most of the atmosphere, there is a considerable wind flow from west to east. The majority of this flow is in the middle latitudes (Lamb, 1982). This circumpolar flow can also be disturbed by variations in atmospheric pressure. These disturbances arise because of differences in the degree of radiant heating of the earth's surface, with lower pressures over cold regions and higher pressures over warm regions, which in turn affect the velocity of the circumpolar vortex. A local increase in pressure will accelerate the flow, whilst a reduction in pressure will cause a deceleration. These processes cause high and low pressure systems to develop, and are in turn moved around the globe by the circumpolar vortex. In general, high pressure systems or anticyclones are maintained along the warm side of the upper wind and low pressure or cyclonic systems occur near the cold side of the main flow.

The pattern of the circumpolar vortex changes (stochastically) from a smooth zonal pattern as described above, to a system of meanders between the middle latitudes and the pole and back to the middle latitudes. This pattern is called meridional and may be long lasting, exerting a blocking effect on the more typical pattern of climate or weather. Meridional flow leads to periods of abnormally extreme weather, with drought in one location and floods in another, and extremes of temperature occurring in the same manner. The summer of 1976 was a striking example, leading to particularly hot and dry weather in the British Isles, but with weather which was wetter than average in Greece and Turkey (Ratcliffe, 1981). Increases in meridional flow and blocking of the atmospheric circulation may have caused or amplified the course of the Little Ice Age during the late sixteenth to early eighteenth centuries. The climate of this period varied considerably over the globe, seen as low precipitation during the Indian monsoons, expansion of the polar ice caps, severe winters in

Europe and yet a warmer climate in Siberia during the seventeenth century (Lamb, 1982).

The effect of these variations in the circumpolar vortex had considerable long-lasting effects on climate and on plant distribution (Laduric, 1971; Grove, 1972; Salinger, 1976; Lamb, 1982). It is clear from Laduric (1971) and Lamb (1982) that not all changes in plant distribution were due to lethal threshold effects; some were also due to non-lethal threshold effects, whereas others may have been competitive (Hyvarinen, 1975, 1976).

The importance of ocean currents to world climate has already been stressed, so any changes in ocean currents may have considerable effects on climate and the distribution of plants. Perhaps the best known periodic variation of ocean currents is the El Niño effect, which is an irregular fluctuation of the ocean currents off the coasts of Peru and Ecuador. The fluctuation has a recurrence time of between 3 and 8 years (Kerr, 1984). This coastline normally has very little precipitation and the adjacent sea is colder than the land because of the dominant, cold Peruvian current. This pattern is generally emphasised by trade winds from the southeast. During an El Niño event, these trade winds are replaced by northerly winds which cause a southward flow of warm equatorial water. The warm water leads to heavy rainfall over the arid regions of Peru (Bjerknes, 1969). Such large changes in the availability of water for plant growth have equally large effects in stimulating the growth of plant species with short life-cycles (W. Hadfield, personal communication).

The time scales of variations in the circumpolar vortex and of the El Niño event are quite short, in the order of years rather than millenia. Kutzbach (1976) has emphasised the importance of intrinsic processes operating on time scales of millenia, with interactions between the atmosphere, the oceans and global ice as the crucial areas for understanding the regular occurrence of Ice Ages. However it is clear that the level of understanding is still in its infancy.

References

Abbot, C.G. (1963). Solar variation and weather. *Smithsonian Miscellaneous Collections*, **146**, No. 3, Washington.

Adhémar, J. (1842). *Révolutions de la Mer*. Paris: Déluges Périodiques.

Bjerknes, J. (1969). Atmospheric teleconnections from the equatorial Pacific. *Monthly Weather Review*, **97**, 163–72.

Black, J.N. (1956). The distribution of solar radiation over the earth's surface. *Archiv für Meteorologie, Geophysik und Bioklimatologie, Series B*, **7**, 165–89.

Bray, J.R. (1968). Glaciation and solar activity since the fifth century BC, and the solar cycle. *Nature*, **220**, 672–4.

Bray, J.R. (1974). Volcanism and glaciation during the past 40 millenia. *Nature*, **252**, 679–80.

Brier, G.W. (1964). The lunar synodical period and precipitation in the United States. *Journal of Atmospheric Sciences*, **21**, 386–95.

Broecker, W.S., Thurber, D.L., Goddard, J., Ku, T.L., Matthews, R.K. & Mesollela, K.J. (1968). Milankovitch hypothesis supported by precise dating of coral reefs and deep-sea sediments. *Science*, **159**, 297–300.

Brooks, C.E.P. (1930). The mean cloudiness over the earth. *Memoirs of the Royal Meteorological Society*, **10**, 127–38.

Bryson, R.A. & Dutton, J.A. (1961). Some aspects of the variance spectra of tree rings and varves. *Annals of the New York Academy of Science*, **95**, 580–604.

Bryson, R.A. & Hare, F.K. (1974) ed. *World Survey of Climatology*, vol. 11. *Climate of North America*. Amsterdam: Elsevier.

Budyko, M.I. (1955). *Atlas Teplovogo Balansa* (Atlas of the Heat Balance). Leningrad: Glav. Geofiz. Observatoria Voeikova.

Budyko, M.I. (1974). *Climate and Life*. New York: Academic Press.

Cadle, R.D., Kiang, C.S. & Lonis, J.F. (1976). The global scale dispersion of the eruptive clouds from major volcanic eruptions. *Journal of Geographical Research*, **81**, 3125–32.

Calder, N. (1974). *The Weather Machine*. London: BBC Publications.

Cox, C.B. & Moore, P.D. (1980). *Biogeography – an Ecological and Evolutionary Approach*, 3rd edn. Oxford: Blackwell Scientific Publications.

Croll, J. (1875). *Climate and Time in their Geological Relations*. New York: Appleton.

Currie, R.G. (1974). Solar cycle signal in surface air temperature. *Journal of Geophysical Research*, **79**, 5657–60.

Davis, O.K. (1984). Multiple thermal maxima during the Holocene. *Science*, **225**, 617–9.

Dewey, E.R. (1960). The 200 year cycle in the length of the sunspot cycle. *Journal of Cycle Research*, **9**, 67–82.

Dyer, A.J. & Hicks, B.B. (1965). Stratospheric transport of volcanic dust inferred from solar radiation measurements. *Nature*, **208**, 131–3.

Eddy, J.A. (1976). The Maunders Minimum. *Science*, **192**, 1189–202.

Gates, D.M. (1980). *Biophysical Ecology*. New York: Springer-Verlag.

Gleissberg, P.A. (1958). The eighty-years sunspot cycle. *Journal of the British Astronomical Association*, **68**, 148–52.

Goreau, T. (1980). Frequency sensitivity of the deep-sea climatic record. *Nature*, **287**, 620–2.

Goudie, A. (1983). *Environmental change*, 2nd edn. Oxford: Clarendon Press.

Grove, J.M. (1972). The incidence of landslides, avalanches and floods in western Norway during the Little Ice Age. *Arctic and Alpine Research*, **4**, 131–8.

Hanel, R.A., Schlachman, B., Rodgers, D. & Vanous, D. (1971). Nimbus 4 Michelson interferometer, *Applied Optics*, **10**, 1376–82.

Hays, J.D., Imbrie, J. & Shackleton, N.J. (1976). Variations in the earth's orbit: pacemaker of the Ice Ages. *Science*, **194**, 1121–32.

Hyvarinen, H. (1975). Absolute and relative pollen diagrams from northernmost Fennoscandia. *Fennia*, **142**, 23pp.

Hyvarinen, H. (1976). Flandrian pollen deposition rates and tree-line history in northern Fennoscandia. *Boreas*, **5**, 163–75.

Imbrie, J. & Imbrie, K.P. (1979). *Ice Ages: Solving the Mystery*. London: Macmillan.

Kelly, P.M. & Sear, C.B. (1984). Climatic impact of explosive volcanic eruptions. *Nature*, **311**, 740–3.

Kerr, R.A. (1984). Slow atmospheric oscillations confirmed. *Science*, **225**, 1010–11.

Kondratyev, K.Y. & Nikolsky, G.A. (1970). Solar radiation and solar activity. *Quarterly Journal of the Royal Meteorological Society*, **96**, 509–22.

Kutzbach, J.E. (1976). The nature of climate and climatic variations. *Quaternary Research*, **6**, 471–80.

Laduric, E. Le R. (1971). *Times of Feast, Times of Famine*. London: Allen & Unwin.

LaMarche, V.C. & Hirschboeck, K.K. (1984). Frost rings in trees as records of major volcanic eruptions. *Nature*, **307**, 121–6.

Lamb, H.H. (1970). Volcanic dust in the atmosphere: with a chronology and assessment of its meteorological significance. *Philosophical Transactions of the Royal Society, Series A*, **266**, 425–533.

Lamb, H.H. (1971). Volcanic activity and climate. *Palaeogeography, Palaeoclimatology, Palaeoecology*, **10**, 203–30.

Lamb, H.H. (1972). *Climate: Present, Past and Future*, vol. 1, *Fundamentals and Climate Now*. London: Methuen.

Lamb, H.H. (1982). *Climate, History and the Modern World*. London: Methuen.

Lockwood, J.G. (1979). *Causes of Climate*. London: Arnold.

McIntosh, D.H. & Thom, A.S. (1978). *Essentials of Meteorology*. London: Wykeham Publications Ltd.

Mesollela, K.J., Matthews, R.K., Broecker, W.S. & Thurber, D.L. (1969). The astronomical theory of climatic change: Barbados data. *Journal of Geology*, **77**, 250–74.

Milankovitch, M. (1920). *Théorie Mathématique des Phenomènes Thermiques Produits par la Radiation Solaire. Gauthier-Villars: Académie Yougoslave des Sciences et des Arts de Zagreb.*

Milankovitch, M. (1930). Mathematische Klimalehre und astronomische theorie der Klimschwankungen. In Handbuch der Klimatologie. 1. Teil A. ed. W. Köppen, W. & R. Geiger. Berlin: Borntraeger.

Milankovitch, M. (1941). *Canon of Insolation and the Ice Age Problem*. Beograd, Köninglich Serbische Akademie. 484pp. (English translation by Israel program for Scientific Translation and published for the U.S. Department of Commerce and the National Science Foundation).

Mitchell, J.M. (1976). An overview of climate variability and its causal mechanisms. *Quaternary Research*, **6**, 481–93.

Monteith, J.L. (1973). *Principles of Environmental Physics*. London: Arnold.

NASA (1971). *Report No. R-351 and SP-8005*.

Oke, T.R. (1978). *Boundary layer climates*. London: Methuen.

Outi, M. (1961). Climatic variations in the north Pacific subtropical zone and solar activity during the past ten centuries. *Bulletin of the Kyoto Gakugei University. B*, **19**, 41–61; **20**, 25–48.

Pearson, R. (1978). *Climate and Evolution*. London: Academic Press.

Pisias, N.G. & Shackleton, N.J. (1984). Modelling the global climate response to orbital forcing and atmospheric carbon dioxide changes. *Nature*, **310**, 757–9.

Plass, G.N. (1956). The carbon dioxide theory of climatic change. *Tellus*, **8**, 140–54.

Ramade, F. (1984). *Ecology of Natural Resources*. (English translation of French edition of 1981) London: Wiley.

Ratcliffe, R.A.S. (1981). Meteorological aspects of the 1975–76 drought in Western Europe. In *Climatic Variations and Variability*, ed. A. Berger & O. Reidel, pp. 355–67. Holland: Dordrecht.

Robinson, N. (ed.) (1966). *Solar radiation*. Amsterdam: Elsevier.

Russell, P.B., Viezee, W., Hake, R.D. & Collis, R.T.H. (1976). Lidar observations of the stratospheric aerosol: California, October 1972 to March 1974. *Quarterly Journal of the Royal Meteorological Society*, **102**, 675–95.

Salinger, M.J. (1976). New Zealand temperatures since 1300 AD. *Nature*, **260**, 310–11.

Schneider, S.H. & Mass, C. (1975). Volcanic dust, sunspots and long-term climate trends: theories in search of verification. *Science*, **190**, 741–60.

Schuurmans, C.J.E. (1981). Solar activity and climate. In *Climate Variations and Variability*, ed. A. Berger, D. Reidel, pp. 559–75. Holland: Dordrecht.

Shackleton, N.J., Hall, M.A., Line, J. & Shuxi, C. (1983). Carbon isotope data in core V19–30 confirm reduced carbon dioxide concentration in the Ice Age atmosphere. *Nature*, **306**, 319–22.

Sonett, C.P. & Suess, H.E. (1984). Correlation of bristlecone pine ring widths with atmosphere ^{14}C variations: a climate–Sun relation. *Nature*, **307**, 141–3.

Sorensen, B. (1979). *Renewable Energy*. London: Academic Press.

Stuiver, M. (1980). Solar variability and climatic change during the current millenium. *Nature*, **268**, 868–71.

Wijmstra, T.A., Hoekstra, S., De Vries, B.J. & Van Der Hammen, T. (1984). A preliminary study of periodicities in percentage curves dated by pollen density. *Acta Botanica Neerlandica*, **33**, 547–57.

4

Climate and vegetation

The purpose of models is not to fit the data but to sharpen the questions.
S. Karlin.

Introduction

The distribution of the world's vegetation types has been known and documented with some degree of accuracy for at least 180 years, dating back to the foundations of plant geography in von Humboldt & Bonpland (1805). Subsequent accounts by von Humboldt (1807), Schouw (1823), Meyen (1846) and de Candolle (1855), for example, soon emphasised the importance of climate in controlling the observed patterns of distribution. The natural result of this connection was a search for the mechanism by which climate could exert this control. At this time (see Chapter 1) many new developments were emerging in plant physiology and Schimper (1898) in particular saw that the climatic control of plant distribution must of necessity operate through basic physiological processes.

This approach to plant geography has remained a chiefly German preoccupation, with considerable contributions by Walter (1931, 1968, 1973, 1976), but with significant contributions in a similar vein by the North American, Cain, (1944), and an inducement to consider the historical perspective (see Chapter 1) by the Russian, Wulff (1943).

The approach adopted by Schimper and by Walter was one of a physiological explanation for why a particular species was able to survive in a particular area. This approach did not generally separate cause and effect and was therefore limited as a predictive tool. If the distribution of global vegetation is to be understood in terms of plant physiology, or ecophysiology, then the starting points should be the known range of physiological responses to climate and the observations of climate. Which optimal match of response and climate, and which type of vegetation possesses these responses, would therefore be the appropriate hypotheses to be tested, with the final predictions being compared with field observations. This approach should also establish the value and direction of ecophysiology and establish new problems for investigation. As Box (1981) describes,

> Predictive modeling [sic], i.e. the rigorous application (extrapolation) of quantitative models to environmental data (at sites other than those used to construct the model) in order to predict actually occurring patterns, can be particularly useful in plant-geography and plant-environment relations, since it provides a ready means of testing the validity of the model and the understanding behind it.

The approaches taken by Holdridge (1947, 1967) and by Box (1981) are representative of an approach to the understanding of the climatic control of plant distribution based on a taxonomy of vegetation and climate. The vegetation of a particular area may be simply classified in overall physiognomic terms (Holdridge), e.g. tropical rain forest, or in more detail (Box) as an overall life form but composed of individuals differing in plant and leaf size, and leaf longevity (less than or greater than 1 year). Although Box's approach has greater detail than that of Holdridge, and also has more climatic correlates, it is no different from that of Holdridge in that vegetation types are predicted by correlation and not from a fundamental physiological basis. However both Holdridge and Box have established that very strong correlations exist between life form or physiognomy and two broad features of climate, temperature and the water budget, measured as the equation of rainfall or precipitation minus evapotranspiration.

An ecophysiological basis

Although the correlation approach fails to establish the mechanisms by which climate may control distribution, it does provide a logical point from which to start. The implications are that the availability of water, or conversely drought, may influence the mass of vegetation, increasing, for example, from an absence of any vegetation in an extreme desert, through a sparse mixture of grasses and trees in rather wetter grassland or savannah, to the dense, galleried structure of a tropical rain forest. The effects of temperature, on the other hand, are multifarious with effects within and beyond the range of plant tolerance (Chapter 2). There are clearly many areas which could be logical starting points in a quest for defining a match between physiological response and local climate; however a massive model including all the known effects of e.g. temperature, water relations, irradiance and wind speed, is neither feasible at present, nor particularly desirable in the present context. What is required is a tautonomous starting point which may subsequently develop and include, step by step, the critical and climatically controlled limitations to the completion and perpetuation of a successful life-cycle.

Solar radiation and growth

A simple but realistic starting point which immediately links climate and plant growth has been initiated by Monteith (1972, 1977). The central thesis is that the growth of a plant is directly related to its ability to intercept solar radiation and to convert the intercepted solar radiation to carbohydrates, or more generally, dry matter. The efficiency of conversion (typically 1.4 g of dry matter, equivalent to about 25 KJ, per megajoule of intercepted radiation, or an efficiency of 2.5%) is rather conservative in character, showing little change with season for those herbaceous species which have been investigated (Monteith & Elston, 1983), although Jarvis & Leverenz (1983) show a seasonal drift in efficiency for *Pinus sylvestris*. However in the absence of further measurements, and for the sake of simplicity, if the efficiency (E) is assumed to change rather little then, over an interval of time t, the accumulated plant weight (w) is related to the incoming solar radiation (S) as,

$$w = E\int_0^t Si\,dt, \qquad (1)$$

where i is the fraction of the incoming radiation which is intercepted by the canopy. Interception is used rather than absorption (a) because it is easier to measure, where

$$a = 1 - r - t, \qquad (2)$$

and r is reflectivity and t is transmissivity:

$$i = 1 - t, \qquad (3)$$

effectively ignoring canopy reflectivity, which may be as low as 5% or 10% of the incoming radiation. In a simple case where all radiant interception is due to leaves and when a constant fraction of radiation penetrates a layer of leaves, the profile of radiation through a plant canopy will be logarithmic and

$$i = 1 - \exp(-KL), \qquad (4)$$

where K is the extinction coefficient for solar radiation (Monsi & Saeki, 1953), and L is the leaf area index, the total area of leaves over a unit area of ground. Accumulated plant weight may therefore be described as,

$$w = E\int_0^t S(1 - \exp(-KL))\,dt, \qquad (5)$$

dependent on leaf area index, the incoming solar radiation and the efficiency of solar radiation to dry-matter conversion. The exponential term in the equation for interception indicates a non-linear relationship between L and w, so that for a typical value of 0.5 for K (Jarvis & Leverenz, 1983) and at a leaf area index of 1, 40% of the incoming radiation is intercepted; this increases to 78% at a leaf area index of 3, 92% at a leaf area index of 5, and 97% at a leaf area index of 7. Therefore changes in L at low mean values will have the greatest effect on w, with small effects once L reaches a value of 5. Although the benefits of an increase in L above 5 will be small in terms of growth, the effect of limiting radiation for regenerating plants beneath such a canopy will be increasingly severe and provides a measure of canopy interference. It also follows that the maximum leaf area index obtainable by a species in non-limiting climatic conditions will also be a result of the ability of the lower leaves to survive in deep shade. Harbinson & Woodward (1984) demonstrated that the photosynthetic rates of the shade leaves of *Fagus sylvatica* and *Ilex aquifolium* were saturated at an irradiance which was only about 1% of the maximum incident on the canopy on a sunny day. Indeed, these leaves were unable to make a photosynthetic advantage from higher irradiance sunflecks. These photosynthetic responses show the species to be shade-tolerant and it is interesting to note from Ellenberg (1978) that shade-tolerant trees, in Europe at least, also cast the densest shade.

Any influences of climate on leaf area index will influence not only the growth of plants but also the ability to cast shade. Temperature exerts a strong effect through its influence on leaf expansion and leaf initiation, both of which increase with temperature to some optimum (e.g. Fukai & Silsbury, 1976; Potter & Jones, 1977; Auld, Dennett & Elston, 1978). Between the low and high temperature thresholds for leaf survival, temperature can be considered as controlling only the rate of leaf expansion rather than the upper or lower limit of leaf area index.

Water relations

The availability of water, on the other hand, has a strong influence on leaf area index. Grier & Running (1977) and Waring *et al.* (1978) have shown clear positive correlations between leaf area index and rainfall in the northwestern coniferous forest of North America. Explanations for the mechanisms underlying this correlation must be sought at a more fundamental level. Expansion of the leaf, for example, is known to be sensitive to changes in plant water potential (Hsiao, 1973). Lockhart (1965), Green, Erickson & Buggy (1971) and Cosgrove (1984) have shown that, at the level of the cell, expansion depends on water intake and is controlled by plant water status. More formally:

$$\frac{dV_w}{V_w dt} = L\,(\Delta\pi - P), \qquad \textbf{(6)}$$

where $dV_w/V_w dt$ is the relative rate of water uptake, L is the hydraulic conductance of the cell, $\Delta\pi$ is the difference in osmotic potential between the inside and the outside of the cell and P is the turgor potential of the cell. The uptake of water into a growing cell is then thought to lead to an irreversible (plastic) expansion of the cell wall:

$$\frac{dV_c}{V_c dt} = m(P - Y), \qquad \textbf{(7)}$$

where the relative rate of irreversible cell wall expansion ($dV_c/\,V_c dt$) is related to P, Y is the yield threshold and m the cell wall extensibility. The yield threshold is the minimum turgor pressure required before any expansion is possible. The cell wall extensibility is a measure of the ease with which the cell wall deforms irreversibly in response to stresses in the wall. **(6)** and **(7)** may then be combined to describe the relative growth rate of the cell dV/Vdt as:

$$\frac{dV}{Vdt} = \frac{mL}{m+L}(\Delta\pi - Y). \qquad \textbf{(8)}$$

Inclusion of such a model as this for cell expansion into a model of whole leaf expansion is more complex because of gradients of water potential across the leaf and spatial variations in active growth (Maksymowych, 1973), in addition to considerations of leaf initiation. However the ease with which the equations for cellular growth are effective in predicting the expansion of leaves (Boyer, 1968 and Hsiao *et al.*, 1976) indicates that such a process is likely to be operative. Tyree & Jarvis (1982) also suggest that the long term integration of growth processes may be related to changes in yield threshold and cell wall extensibility which, in turn, may be controlled by both temperature and by plant water potential. These relationships therefore provide a direct link between the basis of cell extension, the growth of leaves and rate of dry-matter production, or growth of a plant canopy.

It seems likely therefore, when growth is viewed as a physical process of cell wall expansion, that both plant temperature and plant water relations will affect leaf growth through the same pathway. When plant water potential falls to some threshold value during drought, then a different process (leaf abscission) may take place.

McMichael, Jordan & Powell (1973) have shown that premature leaf abscission of cotton (*Gossypium hirsutum*) occurs during periods of drought. When plant water potential only fell to about −0.6 MPa little abscission occurred, but below this threshold abscission increased linearly with the fall in water potential to about 70% abscission at −2.5 MPa. Abscission did not occur during the period of experimentally imposed drought, but occurred after rewatering, and may be controlled by the release of ethylene (Jordan, Morgan & Davenport, 1972).

This control of leaf abscission by plant and soil water potential has been frequently observed for forest trees, and Addicott & Lyons (1973) classified such species as facultatively deciduous, when the leaves are only shed during a period of drought. Facultatively deciduous species are found in many environments of the world. In those areas of the tropical zone with a rather uniform annual climate, the ability of trees to retain leaves throughout the year is the norm (Medway, 1972), whereas leaf abscission during drought can be considered to be by default. *Gossampinus malabarica* and *Tectona grandis* are evergreen in uniformly wet areas but are deciduous in areas with alternating wet and dry seasons (Merrill, 1945). Kozlowski (1976) has reported for the temperate zone that species such as *Ulmus americana*, *Tilia americana*, *Acer platanoides* and *Populus deltoides* show whole or partial defoliation during summers with extreme drought, as may also be observed for *Fagus sylvatica* in Europe. This facultatively deciduous behaviour may also be observed for *Taxodium* (J.J. Ewel, personal communication). However, this response is not apparently the rule for coniferous species and Ewers & Schmid (1981), for example, found that the retention of leaves in conifers increased with aridity.

The massive conifers of the Pacific northwest of North America may store considerable water in the trunk sapwood. As water evaporates from the leaves, water will be taken from the soil and pass along the system to the evaporating surfaces of the leaf. If transpiration exceeds water supply from the roots, as would occur during drought, then water may be taken from sites of storage in the plant. The sapwood may represent a considerable store; Running, Waring & Rydell (1975) calculated that a typical plant of Douglas fir (*Pseudotsuga menziesii*), with a height of 80 m could store 4000 litres (4 m^3) of water, of which 75% may be readily removed during transpiration. A forest stand of Douglas fir was estimated to store about 270 m^3 ha^{-1} of water, which may be removed at 17 m^3 d^{-1} (Waring & Running, 1978). One critical difference between conifers and hardwood species is that in conifers the sapwood may be readily recharged with water after rain whereas for hardwood species (ring porous species in particular), this is not possible. This is because the entire water-conducting column of the hardwood species is disrupted following the withdrawal of water and subsequent cavitation (Siau, 1971). In conifers only individual tracheids are affected, because of the ability of the bordered pits in the tracheid walls to close in response to the pressure gradient that is created, effectively preventing the gas bubbles which are formed from passing to other conducting elements (Gregory & Petty, 1973).

Conifers can therefore use their massive trunks as water reservoirs during drought, and are able to reduce the impact of drought on leaf water potential. In addition, this reservoir increases in volume and in step with increases in leaf area index. The buffering of leaf water potential by water stored in the sapwood has been clearly demonstrated (Waring & Running, 1978; Waring, Whitehead & Jarvis, 1979) and minimises leaf loss during drought.

In the more extreme environments of the alpine timberline and the arctic tree line, there is no evidence that leaf water potential falls sufficiently low to cause leaf abscission (Tranquillini, 1979; Black & Bliss, 1980).

The availability of water for transpiration can therefore directly influence leaf extension and canopy growth. The coincidence of leaf flushing and periods of drought will be critical in this respect. For hardwood species feedback control on leaf area index may be exerted by leaf abscission during severe droughts, while evidence for such a response in conifers appears absent. However, for all trees, the dominant control of leaf area index is through the local evaporative climate (Gholz, Fitz & Waring, 1976; Grier & Running, 1977). Spurr & Barnes (1980) report that for the central areas of the United States of America, an annual rainfall of 380 mm is the lower limit of precipitation to support open woodland, about 500 mm for an open forest and over 640 mm for a closed forest. Further north in Alaska, however, an annual rainfall of as little as 180 mm is sufficient for the development of forest. This suggests that the water budget (precipitation minus evapotranspiration), rather than precipitation alone, may be the critical climatic feature.

Low temperatures

Measurements or predictions of the local water budget can be used to predict canopy development, measured as leaf area index, but can not be used to predict the physiognomy, i.e. whether the local vegetation is needle-leaved coniferous forest or broad-leaved deciduous or evergreen forest, or grassland, shrubland or tundra. The considerations of leaf area index have been confined to the summer growing season; however it is clear for species of conifers which must endure freezing temperatures during the winter that leaf water potentials are likely to be more extreme at that time than during the summer (Kozlowski, 1976). As leaf temperature declines below 0 °C, the water potential also falls, with a coefficient of about 1.2 MPa $°C^{-1}$ (Jones, 1983). Temperature effects, such as changes in membrane structure, the formation of intracellular and extracellular ice and the supercooling of water (Li & Sakai, 1978, 1982) may be as important for survival as drought tolerance, but are also clearly related to the behaviour of water at subzero temperatures. All these features may be crucial to the survival of the characteristic species of a particular physiognomy.

The impact of these low-temperature responses on plant distribution may be readily observed for species which are grown outside their natural geographical range. In Cambridge, UK, for example, minimum air temperatures during the winter of 1981–2 fell to −16.1 °C. Species from warmer winter climates such as *Cistus* and *Hebe* were killed outright, whilst the evergreen oak, *Quercus suber*, suffered complete defoliation. Presumably, if the defoliation of *Quercus suber* was a regular event the species would be less competitive, also requiring more energy to make the leaves which are thicker than the native, but winter deciduous, *Quercus robur*. A regular but facultative winter deciduousness would also limit the actual season for photosynthetic gain. This is typically longer for an evergreen species, although photosynthesis proceeds at a lower rate than for obligate winter deciduous trees in similar climates (Schulze, 1982). Competitive exclusion might therefore limit the poleward spread of a species sensitive to low winter temperatures before low-temperature mortality, which itself would define a precise geographical limit. The implication here is that winter-deciduous species should be more competitive than evergreen species in areas of low winter temperature. That this is likely has been shown in New Zealand, a country where only 4% of the woody species are deciduous (Dumbleton, 1967). The islands of New Zealand have a mild climate because of the moderating influence of the surrounding ocean. However, on the South Island, the Southern Alps have a permanent snow cover and the dense cold air at these high altitudes (3000 m and greater) drains into lowland valleys. The resulting low temperatures have been shown to be fatal for the locally dominant and evergreen *Nothofagus solandri* (Wardle, 1971), leading to a low altitude or 'inverted' timberline. It is only in these areas, where low winter temperatures limit the survival of the evergreen *Nothofagus*, that the deciduous tree *Hoheria glabrata* is dominant (W.G. Lee, personal communication; Wardle, 1977). It fails to spread into the *Nothofagus* forest, although it is well able to survive in the climate (Bussell, 1968*b*), suggesting poor competitive ability. The *Nothofagus*, on the other hand, also fails to dominate *Hoheria* in these areas because of the restricting influence of low winter temperatures, even though killing frosts may not be a regular, annual event (Wardle, 1971).

That low winter temperatures control the facultatively deciduous behaviour of the New Zealand deciduous trees has been demonstrated experimentally (Bussell, 1968*a*) and is suggested by field observations which show the species to be evergreen in warmer areas of New Zealand (Cockayne, 1928). Control of leaf fall in the obligate deciduous species from the northern hemisphere is, however, under photoperiodic control (Wareing, 1956). The actual mechanism leading to low-temperature leaf abscission in the deciduous species of New Zealand is not known, although Zimmerman (1964) has shown that temperatures of only −1 °C or −2 °C are sufficient to cause ice formation in the xylem of the trunk of *Quercus rubra*. This has two powerful effects. First, no water is available to replace that lost by evaporation, even if only through the cuticle, and so frost drought may be a severe problem because the leaf store of water will be small. Waring & Running (1978), for example, have calculated for a plantation of *Pseudotsuga menziesii*, with a leaf area index of 8.4, that the water stored in the foliage was equivalent to less than 0.1% of the total stored in the whole tree. Leaf survival in winter, with frozen xylem vessels, will be strongly dependent not only on enduring low temperatures, but also on frost drought.

The second effect of ice formation in the xylem is the mechanism of recovery following xylem thaw. Once the ice thaws, air bubbles will form and block the vessels. In hardwood species this will inhibit water movement along the pipes of xylem. New xylem vessels must be grown in the spring and these vessels carry by far the most significant volume of water (Huber & Schmidt, 1936). Coniferous species have a distinct advantage here in that water flow through individual tracheids takes place through bordered pits. Just as in the example of cavitation during summer drought described earlier, the pits will serve to isolate the embolism or cavitation resulting from ice formation, so that xylem functioning will be immediate following a thaw (Whitehead & Jarvis 1981).

The similarity between facultative deciduousness during the summer and during the winter has not gone unnoticed by evolutionists. Axelrod (1966), for example, suggests that summer deciduousness was the first to evolve in arid, but warm regions. This ability then facilitated the poleward dispersal of these deciduous species into areas with cold winters, where the species would become winter deciduous. Whether this hypothesis is correct is clearly of interest but not readily demonstrable; what is interesting is the connection between summer and winter deciduousness.

It is not obvious why species which are able to survive the low winter temperatures (of e.g. the boreal zone) are prevented from occurring naturally in warmer climates. Many conifers such as *Larix gmelinii*, *L. laricina*, *Picea glauca* and *Pinus sylvestris* grow well in gardens of the temperate zone, far south of their natural ranges. The suggestion again is that competition will play a key role in limiting their distribution, as has been demonstrated for herbaceous species (Woodward, 1975; Woodward & Pigott, 1975). Certainly many of the trees grow poorly in the warmer climates and some, such as *Picea mariana* and *Populus balsamifera*, become much shorter lived.

Although the control of both the northerly and the southerly limits to the distribution of a major physiognomic type may lie in competitive relationships with other species, as a mechanistic explanation it leaves a lot to be desired. Competitive ability can be simply described as maximal dry-weight accumulation, which is dependent on canopy leaf area index and the extinction coefficient for solar radiation (**5**). The most competitive individuals would therefore have the highest product of leaf area index and extinction coefficient, or interception. Reduced interception, resulting from leaf abscission, is an obvious effect of drought but appears less likely as a response to low but not lethal temperatures.

Levitt (1980), in reviewing observations on plant responses to low temperatures, including freezing, lists a range of fundamental biochemical changes which are energy requiring. These include increases in unsaturated lipids, the accumulation of water soluble solutes such as sugars and sugar alcohols, the accumulation of membrane proteins, increases in proteins for scavenging superoxides and peroxides (Nakagawara & Sagisaka, 1984), and even an increase in ATP which has proved to be an effective cryoprotectant (Santarius, 1984). It is clear that the cell membranes are one of the primary sites of injury at low temperatures. The membrane may be protected by an unspecific colligative solute action, resulting from the increase of solutes (cryoprotectants). The hypothesis is that an increase in the cryoprotectant concentration will lower the critical activity of an injurious solute in the cell (Meryman, Williams & Douglas, 1977). It has also been suggested that the areas of the cell membrane which are particularly at risk from freezing stress may be induced into conformational changes, which produces more resistant membranes (Volger & Heber, 1975). Failure to undergo conformational changes may lead to leakage of ions and amino acids (Ona & Murata, 1981), quickly followed by cell death. It appears, therefore, that the cell membrane is the primary site of both chilling (temperatures greater than 0 °C and less than about 10 °C) and freezing injury (Heber *et al.*, 1981) and is a site which must be protected. However, the mechanisms of hardening, the slow change in ability to withstand low temperatures, and the sites of damage are still unclear. Equally uncertain are the mechanisms involved in chilling sensitivity and resistance. It has been proposed that cells survive chilling temperatures by retaining a membrane phase which is fluid (liquid crystal). When the phase changes at some threshold temperature to a more solid gel state, then membrane leakage and probably death follows. It has also been suggested that increasing the unsaturated fatty acid content of the membrane lowers the temperature at which the phase change occurs, increasing the resistance to low temperatures (Lyons, Raison & Steponkus, 1979*b*).

Frost drought

When low temperatures cause freezing in the leaf, extracellular ice forms. The cause of the spatial organisation of freezing is unclear but is presumably related in part to gradients of osmotic potential. The appearance of intracellular ice is thought always to be fatal (Levitt, 1980). The formation of ice removes liquid water from the leaf and so the solute concentration increases. In addition, the vapour pressure over ice is lower than over liquid water at the same temperature. These two properties ensure further movement of water from the cell to the site of extracellular-ice formation and a subsequent tendency for the cell to contract. The proportion of the total water in the leaf which remains unfrozen decreases hyperbolically below the temperature at which extracellular ice is just formed (Gusta, Burke & Kapoor, 1975). So for a typical example with a cell osmotic potential of −1.8 MPa, 40% of the original water will be unfrozen at −5 °C, 20% at −10 °C and 10% at −25 °C. An important feature associated with the successive diminution of the liquid water content is the increase in solute concentration. As Franks (1983) pointed out, the concentration factor will also depend on the initial osmotic potential, increasing with increasing osmotic potential. Increases in solute concentration will be accompanied by large changes in pH and in ionic strength, leading to irreversible denaturation of proteins and other macromolecular structures. It follows that freezing injury could be a significant end result.

Frost drought is clearly an appreciable problem for plant survival. Some species avoid freezing by the supercooling of water below the freezing point that is predicted from water potential. Supercooling is the rule for most species, which differ rather in their degree of supercooling (Burke *et al.*, 1976). When cooled moderately rapidly, pure water will fail to freeze at 0 °C; this is because the process of freezing must be preceded by nucleation. A nucleus for freezing is a cluster of water molecules produced at random, in which the spacings and orientations are so like ice that water condenses on the cluster surface and freezing occurs (Franks, Mathias & Trafford, 1984). The formation of the clusters is enhanced by particulate matter and low temperature (Franks, 1982). Pure, supercooled water nucleates spontaneously *in vitro* at about −39 or −40 °C, suggesting a lower limit to the survival of those species which are solely dependent on supercooling as a mechanism of avoiding intracellular freezing. However, Rasmussen & MacKenzie (1972) have shown that the accumulation of solutes can reduce the temperature of spontaneous nucleation below −40 °C. Supercooling to temperatures of at least −70 °C has been observed in the buds, and in some cases the leaves, of various hardy conifers such as *Picea glauca*, *Larix laricina* and *Larix sibirica* (Sakai, 1979, 1983). Franks (1983) has suggested that supercooled water is physically separated from the freezing domain, as in the case of the intra- and extracellular domain, and that this water is not subject to osmotic dehydration. This would be an effective mechanism of avoiding frost drought but totally dependent on the reliability of supercooling. Cells would still, of course, need to survive very low temperatures. Sakai (1983), however, says that the survival of the winter buds of conifers at low temperatures is due both to supercooling of water in the primordial bud, and to extraorgan freezing (the formation of ice in the bud scales). The formation of ice in the scales near the bud will lead to a withdrawal of liquid water from the bud to the scales, down a gradient of water potential, leading to frost drought. Sakai has shown that the presence of a parenchymatous pith cavity beneath the crown of the primordial bud of species of *Picea*, *Abies* and *Larix* effectively prevents ice nucleation spreading into the bud along the xylem. For these species, the presence of a pool of supercooled water will ensure the presence of some water with low solute concentrations, reducing frost drought.

Although the mechanisms for surviving low temperatures are still not known with any certainty, it is clear that the ability to survive temperatures of 10 °C and less is dependent on a range of energy requiring processes, many of which are concerned with protecting the integrity of cell membranes. When the range of plant responses to low temperatures is investigated then particular cardinal temperatures, or ranges of temperature, may be recognised.

The temperatures over which chilling sensitivity is commonly observed lie between −1 °C (without freezing) and +12 °C (Larcher & Bauer, 1981). It is likely that species may be able to control the threshold for chilling resistance by changes in membrane structure.

Once temperatures fall below 0 °C, freezing of water within the plant becomes an increasingly likely event. However, the next and obvious cardinal point which emerges is the temperature at which supercooled water nucleates spontaneously to form ice, with a limit, *in vitro* at least, of −39 or −40 °C. Becwar & Burke (1982) provide a number of examples of woody species which are killed once the stems are cooled within the range of −35 to −40 °C. Their experimental evidence suggests that (*in vivo*) rather small volumes of deeply supercooled water, perhaps in buds, freeze at the temperature of spontaneous ice nucleation, with death the inevitable result.

Low temperatures and plant distribution

The ability of a species to survive low winter temperatures to a certain threshold has been an attractive starting point for describing the climatic control of plant distribution. Raison *et al.* (1979) describe a clear correlation between the temperature at which the cell membrane changes from the liquid-crystalline to the gel state and the geographical range of a number of species. The implied hypotheses are that irreversible cell injury occurs, probably because of a reduced permeability to water diffusion once the membrane is in the gel state, and that the ability to resist this change in state at low temperatures is genetically controlled, setting a finite limit to plant survival at low temperatures. Temperature is strongly correlated with latitude, with minimum temperatures decreasing in a poleward direction (Larcher & Bauer, 1981), therefore establishing a strong correlation between the proposed cause and effect. The fact that many herbaceous and chilling sensitive plants are also important crop plants has also meant that many inadvertent experiments have tested the proposed hypotheses. Lyons, Graham & Raison (1979*a*) and Levitt (1980) described the chilling sensitivity of a number of crop plants, a sensitivity which will on occasion be tested perhaps during extreme temperatures in the normal range of cultivation (Lamb, 1982). That the threshold for chilling injury may be selected, probably in response to local climatic conditions, has been shown by Patterson, Paull & Smillie (1978), who demonstrated a decrease in chilling sensitivity with altitude in the Andes of South America for populations of the tomato, *Lycopersicon hirsutum*.

The critical temperatures for leaf survival may be the lowest in winter for species with wintergreen leaves, even though the leaves will be at their most resistant due to hardening. Less extreme low temperatures during the period of maximum growth and in the dehardened condition may also prove to be effective in limiting plant distribution (Lyons *et al.*, 1979*a*).

Moving in a poleward direction, from low altitudes in the tropics, will be associated with periods of the year in which the minimum temperature falls below the chilling range into the subzero, or freezing, range. The earlier discussions on the effects of freezing temperatures on plants have shown that both low temperatures *per se* and frost drought may be critical in determining plant survival and in turn geographical spread. Gusta *et al.* (1975) have shown that the freezing resistance of a range of cereals is closely correlated with the tolerance of diminishing quantities of liquid water at freezing temperatures. Later studies by Rajashekar, Gusta & Burke (1979) and Gusta, Fowler & Tyler (1982) maintain that as most of the freezable water is indeed frozen at −10 °C, there will be little change in this component or in frost drought at lower temperatures, therefore providing no explanation of differing lethal temperature thresholds such as for Kharkov winter wheat at −17 °C, Puma rye at −30 °C and Kentucky bluegrass (*Poa pratensis*) at −38 °C. As with chilling sensitivity, they maintain that it is the temperature at which the plasma membrane undergoes a phase transition which determines the critical temperature for injury. Below the temperature at which the phase transition occurs, the plasma membrane ceases to separate the extra- and intracellular domain of water, possibly allowing the formation of intracellular ice, but certainly leading to irreversible and lethal damage to the cell.

It emerges, therefore, that both frost drought and membrane sensitivity to low temperatures are likely to be the mechanisms which control plant survival and therefore distribution at freezing temperatures. Sakai (1978) has shown clearly that the distribution of dominant species of trees in Japan is strongly correlated with their winter freezing resistance. Broad-leaved, evergreen trees dominate the warm, southern regions of Japan, where the minimum temperature rarely falls below about −15 °C. The frost resistance of these species correlates strongly with the trend of the isotherms for extreme minimum temperatures. The region dominated by evergreen trees rapidly changes to a region dominated by broad-leaved deciduous trees, once the minimum temperature drops below about −15 °C. On the most northerly of the Japanese islands, Hokkaido, where minimum temperatures may fall below −30 °C, a mixed forest of broad-leaved deciduous and coniferous species has developed.

This work of Sakai, and conclusions drawn by de Candolle (1855), Parker (1963), Sakai & Weiser (1973), George *et al.* (1974) and Larcher

& Bauer (1981), all suggest that the poleward spread of a species or a physiognomic type of vegetation is controlled by the minimum temperature which will be regularly experienced.

Larcher & Bauer (1981) have constructed a world map describing those areas where different low-temperature thresholds will control plant and vegetation distribution. If the hypothesis that winter temperatures can control vegetation distribution is to be valid and testable then the obvious starting point is the prediction map such as that described by Larcher & Bauer. Given a map of predictions, then the hypothesis may be tested, initially by comparisons with well-established, geographical distributions. Further testing would involve analyses of the mechanisms such as already have been described. The aim of this chapter is to establish such a map based on those critical cardinal temperatures which have emerged from experimental analyses. Such an approach has been applied by Sakai & Weiser (1973) to the distribution of forest trees in North America.

The maximum frost resistance for the leaves of 128 species of evergreen broad-leaved trees and 130 species of coniferous trees are shown on Figs. 4.1(*a*) and 4.1(*b*). These data, and those for the frost resistance of overwintering buds, have been taken from a number of sources, all of which have broadly employed the experimental techniques described by Sakai & Weiser (1973) and Sakai (1978) and are from Kaku (1971), Rajashekar & Burke (1978), Sakai & Wardle (1978), Larcher (1981*a*), Larcher & Bauer (1981), Larcher & Winter (1981), Becwar & Burke (1982), George (1982), Larcher (1982), Oohata & Sakai (1982), Paton (1982) and Yoshie & Sakai (1982).

It emerges from Fig. 4.1(*a*) that the frost resistance of the leaves of evergreen broad-leaved trees is very restricted, with 91% of all observations falling within the range of 0 to −15 °C. However, the causes of death, whether frost drought, low temperature effects on the cell membrane or the limit of supercooling, are yet to be established. The leaves of the conifers show a much wider range of resistance and include some of the broad-leaved but sensitive species of *Agathis* and *Araucaria* from the warm, evergreen forests of Australia and New Zealand. The broad range of frost resistance comes to a rather abrupt end at about −40 °C, which may coincide mechanistically with the spontaneous nucleation of supercooled water. A number of species have no temperature thresholds, surviving the lowest experimental temperatures (typically −70 to −80 °C).

Deciduous broad-leaved species are not included in Fig. 4.1(*a*) because of leaf abscission in the winter. However, the buds of these and the broad-leaved evergreen species must survive the winter cold. It is interesting

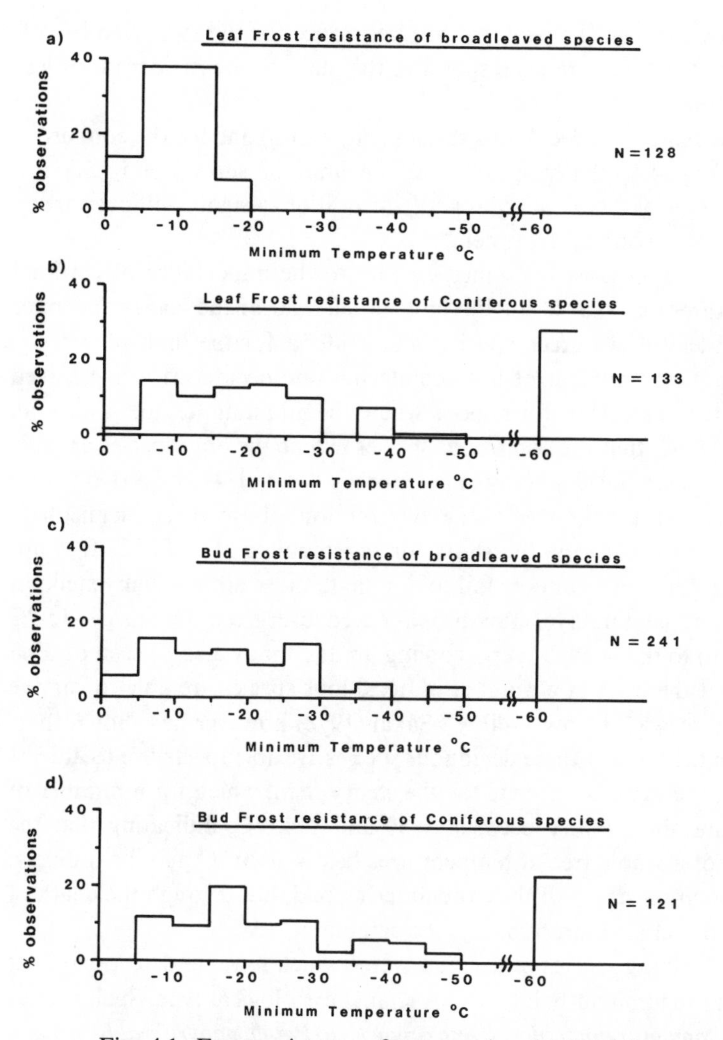

Fig. 4.1. Frost resistance of tree species. (*a*) Leaf frost resistance of broadleaved species. (*b*) Leaf frost resistance of coniferous species. (*c*) Bud frost resistance of broad-leaved species. (*d*) Bud frost resistance of coniferous species.

to note that there are no obvious distinctions between the histograms of bud frost resistance for either the broad-leaved (Fig. 4.1(*c*)) or coniferous (Fig. 4.1(*d*)) species, which are in turn very similar to the resistance of coniferous leaves (Fig. 4.1(*b*)). Sakai & Wardle (1978) and Sakai (1978) show very strong correlations between the threshold temperatures for leaf and bud resistance amongst the evergreen broad-leaved species and for the coniferous species. The peaks of bud resistance for the evergreen broad-leaved species therefore correspond in the main to the histogram of leaf frost resistance.

For the remaining deciduous species (Fig. 4.1(*c*)) and for the coniferous species (Fig. 4.1(*d*)) it appears that the limit of supercooling *in vitro* emerges again as a critical point with, in addition, a significant proportion of species surviving all treatments.

These data on frost resistance emphasize the importance of cardinal temperatures at −15 °C for the limit of survival of the leaves and buds of broad-leaved evergreen species, and −40 °C for the limit of survival for a significant number of both coniferous and deciduous broad-leaved species. It is also clear, for regions where the minimum temperatures fall below − 15 °C, that the coniferous species will be the only species capable of an evergreen physiognomy. Yet George *et al.* (1974) and Sakai (1978) have ascribed the limits to the distribution of northern hemisphere deciduous forests to the threshold for supercooling at −40 °C. It is not clear therefore, why conifers fail to dominate those areas which extend at least from the limits of the broad-leaved evergreen forests (−15 °C minimum) to the −40 °C supercooling limits. It may also be seen on Fig. 4.1(*c*) that the buds of a number of deciduous species are able to survive temperatures well below −40 °C. Sakai (1978) however has shown that, although the buds of these deciduous species are able to survive to at least −70 °C, the same is not true for the stem xylem, which for a number of species sustained injury between −30 and −40 °C, indicating that the survival of a whole tree at temperatures below −40 °C may be unlikely, not through the death of the overwintering bud, but through the death of the xylem and also, presumably, the cambium.

The remaining species of deciduous tree which have no detectable frost limit fall predominantly into a characteristic ecological type. Such species include *Populus tremuloides*, *P. maximowiczii*, *Betula papyrifera*, *B. ermanii* *B. platyphylla*, *Salix nigra* and *S. sachalinensis* (Sakai & Weiser, 1973; Sakai, 1978), and are all pioneer species, incapable of surviving in shade and with rather short lifespans. This relationship is not the case for coniferous species. Shade-intolerant species will not create and dominate a long-lived forest, but are likely to be replaced by slower-growing, longer-lived and more shade-tolerant conifers of the boreal zone, such as *Picea mariana* (Spurr & Barnes, 1980). It follows, therefore, that the majority of potentially dominant and deciduous trees have low-temperature thresholds for winter survival in the range of about −15 to −40 °C.

The data and interpretation of Sakai & Weiser (1973), George *et al.* (1974) and Sakai (1978) suggest that deciduous forests should dominate those latitudes where the minimum temperature does not fall below −40 °C. The data presented on Fig. 4.1 would also suggest a higher temperature limit of about −15 °C for the equatorial spread of the deciduous forest. Where the minimum temperature is greater than −15 °C, leaf survival of many broad-leaved, evergreen species will not be limited by minimum temperatures and the species will be continually in leaf. For example, in the broad-leaved, evergreen forests of Japan, Kusumoto (1957) has shown that the dominant evergreen species such as *Castanopsis sieboldii* and *Cinnamomum camphora* are able to make significant photosynthetic gains of dry weight throughout the year, albeit with rather reduced rates in the cold winter months of the year. Deciduous species such as *Quercus acutissima* and *Q. serrata*, which make no photosynthetic gains during the winter, are often pioneer species in cleared areas and are naturally excluded by the native broad-leaved and evergreen species (Satoo, 1983).

Kira & Shidei (1967) suggest that growth and, in effect, competitive ability in forests are related to the product of leaf area index and the growing season. The broad-leaved evergreen forests with high leaf area indices, ranging from about 5 to 10, and with an active growing season of at least 8 months (Satoo, 1983), will therefore be more productive and competitive than deciduous species, with leaf area indices rarely greater than 5 (Jarvis & Leverenz, 1983) and growing seasons, in Japan, of about 6 months. Other physiological features of evergreen species may also be critical in this competitive balance. Chabot & Hicks (1982), in their review on the ecology of leaf lifespan, conclude, for example, that long-lived leaves make more efficient use of soil nutrients than deciduous species.

The little evidence that exists for the competitive relationships between coniferous species of the boreal zone and broad-leaved, deciduous trees of the temperate zone suggests that the same pattern of competitive exclusion as described earlier will prevail. Olsen (1958) describes succession on sand dunes from dune grasses, through forests of *Pinus banksiana* and *Pinus strobus* to domination by *Quercus velutina* on the southern margin of Lake Michigan in North America. Botkin, Janak & Wallis (1972) used a number of physiological features of forest trees, in particular rates of growth, evapotranspiration and longevity, to predict by modelling the species dominance in the Hubbard Brook area of northeastern North America. At low altitudes species such as *Acer saccharum*, *Fagus grandiflora* and *Betula alleghaniensis* were predicted to be dominant, fitting closely with observation. At cooler altitudes, above about 800 m, predictions and observations show an increased importance of evergreen coniferous species such as *Picea rubens* and *Abies balsamea*.

Palaeoecological evidence, such as that described in Chapter 1, suggests that for example, Ohio, (North America) which is at present an area of deciduous forest, was originally dominated by boreal, coniferous genera such as *Picea* and *Larix* during the colder, immediately post-glacial period at 14 000 years BP. These species were apparently excluded by deciduous genera such as *Ulmus*, *Quercus*, *Acer*, *Carya* and *Fagus* during significantly warmer intervals of climate (Davis, 1980).

This same pattern of a reduced geographical range of boreal conifers in the temperate zone, with time, since the last ice age, has also been shown in the British Isles for *Pinus sylvestris*. Bennett (1984) has shown on the basis of macrofossil and pollen evidence that *Pinus* first spread from continental Europe into the south of England at about 10 000 BP, during the period of climatic amelioration following the last glaciation. *Pinus* reached central Scotland by about 4000 BP. However between 10 000 and 4000 BP most of the English and Irish range of *Pinus* was eliminated, being replaced by *Corylus*, *Ulmus*, *Quercus* and *Alnus* in a range of habitats, and also by the spread of blanket bog. The implication, but not demonstration, from this study is that the more productive deciduous species out-competed *Pinus*. However, the suggestion of the importance of the spread of blanket bog in the demise of *Pinus* also implies a direct climatic limitation to its spread (Bennett, 1984).

Geographical limits and climate

There is abundant evidence for the climatic limitation of geographical spread, particularly when described in terms of the low-temperature limits for survival. As a general rule it can be concluded that the poleward spread of a particular physiognomic type of vegetation will be strongly controlled by minimum temperature and the physiological ability to survive low temperatures. Larcher (1982) views this low temperature limit as a potent agent of natural selection. What is less clear is the cause of the geographical limit of a physiognomic type in the opposite, equatorial direction. Temperatures at similar altitudes will generally increase, but insufficiently for high temperatures to be a severe limiting factor (Larcher, 1980). The evidence suggests that, at the species level, competitive exclusion is likely to be the key. However the boundary generated in this way is likely to be strongly dependent on a range of local features of climate, soil, species composition and growth. It would be expected therefore that the geographical limits to survival in the equatorial direction may be less precise than in the poleward direction. The absence of a direct climatic limitation to survival, at least in terms of low temperature, will allow the species to grow and complete a normal

Table 4.1 *Cardinal minimum temperatures and expected dominant physiognomy*

Temperature range (°C)	Phenomenon	Expected physiognomy
>15	Temperature not limiting	Broad-leaved evergreen when rainfall adequate
−1 to 15	Chilling	Broad-leaved evergreen when rainfall adequate
−15 to 0	Freezing and supercooling	Broad-leaved evergreen
−40 to −15	Freezing and supercooling	Broad-leaved deciduous
< −40	Freezing and supercooling	Evergreen and deciduous needle-leaved (coniferous)

life-cycle, as indeed may be seen for many boreal conifers in temperate zones. Similarly, and in native vegetation, Satoo (1983) shows the presence of a small component of deciduous species in an otherwise predominantly broad-leaved evergreen forest in Japan.

This imperfect boundary, perhaps resulting from poor competitive abilities, is assured, at least in part if not completely, by the physiological commitment of these species to low-temperature tolerance. The earlier review on tolerance of low temperatures and frost drought describes the metabolically active manner in which a species maintains a tolerance, by stimulating or diverting metabolism towards protection rather than growth. Low-temperature survival clearly has a cost seen as a reduced competitive ability, particularly in regions where survival in less extreme conditions is required.

It is possible, given these conclusions and the range of published information reviewed earlier, to predict the geographical extent of a major physiognomic type of vegetation based on measurements of minimum temperature. These basic predictions are shown in Table 4.1 for five cardinal ranges of temperature. The first range is for areas in which the minimum temperature is always greater than the upper threshold temperature for chilling injury. No biochemical responses would be expected to cope with non-critical low temperatures, and the most productive and competitive physiognomic type of vegetation in this climatic province should be broad-leaved and evergreen (Cannell, 1982). Should rainfall be inadequate to fund annual evapotranspiration then it is expected that the lack of any significant low temperature tolerance, in what will be tropical or subtropical vegetation, would be associated with a drought deciduous vegetation (Axelrod, 1966; Doley, 1981).

In the direction of increasing low-temperature tolerance, or survival, drought tolerance is associated with vegetation which is frost tolerant but with no strong evidence for this in chilling-tolerant vegetation (Lyons *et al.*, 1979*a*, *b*; Patterson, Graham & Paull, 1979). However the significant changes in physiology associated with changing the threshold of the membrane phase transition has a cost in terms of energy requirement and reduced physiological efficiency (Lyons, 1973; Quinn & Williams, 1978). Chilling-sensitive vegetation will be effectively excluded from areas with annual frost because of frost sensitivity but excluded competitively, and probably imperfectly, from areas in which temperatures do not fall into the chilling range.

Broad-leaved and evergreen vegetation is able to survive minimum temperatures to about −15 °C. The precise physiological nature of this limitation is unclear. This ability is often, but not necessarily, associated with an ability to survive both frost drought and summer drought (Levitt, 1980), resulting from structural changes in the cell membrane, increases in cryoprotectants, and various other changes, such as in cell wall thickness and leaf morphology (Paleg & Aspinall, 1981). This frost-resistant vegetation should show a sharply defined geographical limit associated with minimum temperatures of −15 °C, but with a less well-defined limit in areas with chilling but not freezing temperatures. It is also likely that some variation in frost resistance may occur along a climatic gradient, with selection pressure for a more competitive and less frost-resistant population in warmer climates (Larcher, 1981*b*).

When the minimum temperature falls below about −15 °C, then the characteristic vegetation should also be broad-leaved but winter deciduous. Leaves will only develop in the warm growing season and need no extreme frost tolerance, except the ability to endure slight spring frosts. This type of vegetation will clearly throw little shade in the winter and may be unable to prevent the invasion of frost-tolerant, needle-leaved species, which may also be capable of photosynthetic gains in the winter (Schulze, 1982). However the high rates of both photosynthesis and leaf growth are potent competitive features of deciduous vegetation (Schulze, 1982). It is also clear that conifers are incapable of significant photosynthesis during cold winters, although the onset of photosynthesis may be earlier than for deciduous species in the spring (Pisek & Winkler, 1958).

For late successional species of deciduous trees it appears that the temperatures of the dormant buds, xylem or cambium are critical for survival, with limits falling between −15 and −40 °C (Sakai and Weiser, 1973; George *et al.*, 1974). This relationship is evidently not true for the early pioneer and short-lived genera of *Betula* and *Populus* with low temperature tolerances exceeding −40 °C and often equalling those of boreal conifers (Sakai and Weiser, 1973). The nature of the relationship between successional status and frost tolerance remains to be defined.

Frost, tolerant conifers will be able to spread into the broad-leaved and deciduous zone to a degree dependent on competitive interactions between the species of the two physiognomic types. The conifers also undergo wide-ranging and significant changes in leaf and bud physiology during the period of hardening before winter and during winter itself. Such changes occur in the chloroplasts (Senser *et al.*, 1975; Senser and Beck, 1977), in the cell membranes (Ziegler and Kandler, 1980) and in the osmotic properties of the cell (Senser *et al.*, 1971; Larcher *et al.*, 1973). The leaves and buds also become extremely tolerant of frost drought (Sakai, 1979, 1983).

The hardy conifers therefore fit into an overall scheme in which increased frost tolerance and therefore geographical range in the poleward direction requires extensive modifications in cell biochemistry, changes which must take place during the growing season, before the onset of winter. In contrast, this poleward spread will be at the expense of equatorial spread because of the reduced competitive ability due, at least in part, to the extensive biochemical changes involved in winter hardiness.

Predictive model for geographical distribution

The earlier discussions have served to provide a scheme with the potential for describing the geographical range of major physiognomic types of vegetation on the basis of the annual minimum temperature. In addition, it also emerges that the leaf mass, measured as leaf area index, will be under the control of the local hydrological budget, with a low leaf area index being expected in a dry environment and the converse in a wet environment.

In dry environments there will be competition between the typically deep-rooted trees and the more shallow-rooted shrubs and herbaceous species, for the limited downflow of rainwater. Walter (1973) has suggested, for the arid savannah-like lands, that there is a rather stable mixture of grasses and trees. The shallow-rooted grasses have first call on percolating rainwater, whilst the deep-rooted trees take up water both from the upper horizons of the soil and from the deeper subsoil water. The subsoil water is derived in the main from local rainfall and results from variations in drainage patterns (Bate, Furniss & Pendle, 1982; Walker & Noy-Meir, 1982). In these areas the grasses may be considered to be the dominant species through their influence on the percolation of water through the soil. The trees on the other hand, in climates where rainfall is limiting, can be viewed as subordinate invaders. The leaf area index of the woody species in savannah is usually about 1 or less, perhaps reaching a maximum of 3 (Menaut & Cesar, 1982; Jarvis & Leverenz, 1983).

Taking the lead provided by these considerations of the ecology of the savannah regions, it is proposed that where the local hydrological budget is sufficient to maintain a leaf area index of 1, then grassland should be the dominant physiognomic type. The grassland would then grade into vegetation dominated by shrubs up to a leaf area index of about 3, above which trees should be the dominant form. A pictorial version of this scheme is presented in Walter (1939).

In summary, it is proposed that leaf area index and vegetation structure and mass can be predicted from the hydrological budget, whilst the life form of the vegetation may be predicted from minumum temperatures. In addition, if monthly predictions of the hydrological budget may be achieved, then seasonal drought may be predicted, which in turn may influence physiognomy, particularly in climates where the minimum temperature is greater than freezing point. In these areas it is expected that the vegetation should be drought deciduous, if the drought is severe (Axelrod, 1966; Doley, 1981).

Model for hydrological balance

The aim of this model is to predict the total annual evapotranspiration from canopies with a range in leaf area indices (L) and in local climatic conditions. The predicted leaf area index at a site will be that for which evapotranspiration does not exceed precipitation to the extent that drought occurs, leading to leaf abscission. The method of achieving this prediction will emerge from further discussions of the model. The model must use basic meteorological records, which are the only readily available data from much of the world's surface (e.g. Müller, 1982 and references therein), for estimating transpiration.

Evaporation from any wet surface is a function of the energy absorbed by the surface and the gradient of water vapour concentration between the surface and the air above. The majority of the net radiant balance of a surface is dissipated as sensible (convection) and latent (evaporation) heat, with much smaller amounts of energy being stored as heat (changing the temperature) or as the products of a photochemical reaction (photosynthesis). The absorbed energy is from the sun, as short-wave or solar radiation, and from terrestrial objects, as long-wave or infra-red radiation. As energy is neither created nor destroyed, it follows that the energy gained by a surface must be balanced by the energy lost.

The major radiative properties of a forest canopy are described in Fig.

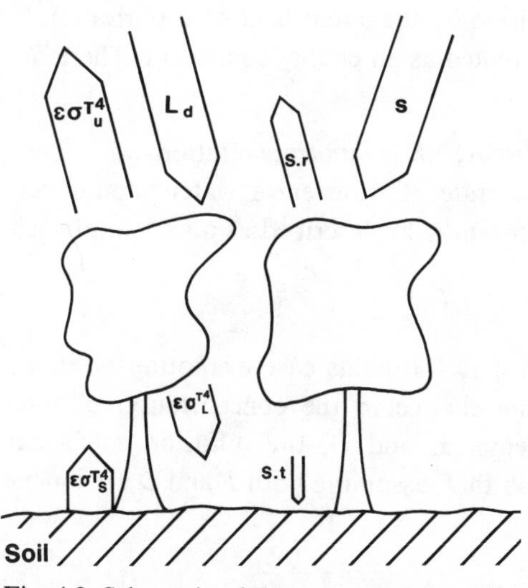

Fig. 4.2. Schematic of the energy balance of a tree canopy.

4.2. The canopy absorbs solar radiation (S) and an atmospheric downward flux of long-wave radiation (L_d). The canopy also gains an upward flux of radiation from the soil surface, with a flux described by the Stefan–Boltzmann equation:

$$L = \varepsilon\sigma T_s^4, \quad \textbf{(9)}$$

where ε is the long-wave emissivity of the soil, T_s its absolute temperature, and σ is the Stefan–Boltzmann constant ($5.67 \times 10^{-8} Wm^{-2} K^{-4}$).

The temperature of the lower surface of the canopy (T_l) and the soil surface will be very similar, and so the upward long-wave flux will be balanced by a downward flux from the canopy ($\varepsilon\sigma T_l^4$). Similarly the net exchange of long-wave radiation between adjacent trees, particularly over a long period, may be rather small and is therefore ignored.

The temperature of the upper surface of the canopy rises by the absorption of radiation. The surface also emits radiation towards the atmosphere, with a flux determined by the surface temperature, T_u, as $\varepsilon\sigma T_u^4$. Some solar radiation fails to be absorbed by the canopy because of reflection (r) and transmission (t) to the soil surface. The total energy balance, or net radiant balance (R), of the canopy may therefore be described:

$$R = (Ld+S)-(Sr+St+\varepsilon\sigma T_u^4). \quad \textbf{(10)}$$

The net radiant balance is in the main dissipated by evapotranspiration (E) and by convection (C). Evapotranspiration may be considered as a loss in weight of water (E), and the energy required to evaporate this water is obtained on multiplication by the latent heat of vaporisation (λ) so that evapotranspiration is treated as an energy equivalent. Therefore,

$$R = \lambda E + C, \tag{11}$$

where the aim is to estimate λE (evapotranspiration).

In general terms, the rate of transfer of water vapour between the canopy and the air above may be described as a variant of Fick's Law:

$$F = -D\frac{dC}{dz}, \tag{12}$$

where the rate of transfer, F (in this case evapotranspiration, E), is a function of dC/dz, the change in the concentration of the diffusing substance, C, with height z, and D, the diffusion coefficient. (**12**) is commonly integrated so that, assuming both F and D are independent of z, then

$$F = \frac{C_{z1} - C_{z2}}{(z_2 - z_1)/D}, \tag{13}$$

where $C_{z1} - C_{z2}$ is the concentration difference over the distance z_1 to z_2 from some position (Woodward & Sheehy, 1983). For diffusion across the leaf surface (**13**) can be modified to

$$E = (\chi_c - \chi_a)/r, \tag{14}$$

where χ_c is the absolute humidity of air in the leaves of the canopy (gm^{-3}), χ_a the absolute humidity of the air and r the resistance to gaseous diffusion (typically in units of sm^{-1}).

The local climate will influence evapotranspiration through a number of changes in canopy temperature which, in turn, affects the (saturated) absolute humidity within leaves. In addition, changes in the absolute humidity of the air and direct effects on the resistance to diffusion will also influence evapotranspiration. The resistance to diffusion is complex but includes the resistance of stomata to water loss, effected by variations in stomatal aperture and which, by direct analogy with Ohm's law, are in series with the aerodynamic or boundary layer resistance to diffusion. The boundary layer is the depth of air above a surface through which diffusion is retarded by the presence of the surface (Monteith, 1973). The boundary layer resistance is strongly influenced by wind speed and by leaf size (Gates & Papian, 1971; Grace, 1977). The stomatal resistance is also dependent on climatic features, in particular irradiance, temperature, vapour pressure deficit and plant water potential (Jarvis, 1976).

Penman (1948) derived an equation which included the radiative and diffusive aspects of evaporation, based on generally observed meteorological variables, and estimated the potential evaporation of a site. The potential evaporation is the predicted evaporation from an extensive free water surface, and as such is imprecise for canopies differing, for example, in leaf area index and in stomatal characteristics. Monteith (1965) modified the equation so that it was more appropriate for plant canopies. In general terms the Penman and Monteith equations rearrange (**11**) in terms of evapotranspiration. In more specific terms:

$$\lambda E = \frac{sR + \rho c_p [e_s(T_a) - e]/r_{aH}}{s + \gamma(r_a + r_s)/r_{aH}}, \qquad \textbf{(15)}$$

where R is the net radiant balance of the canopy, ρ the density of the air, c_p the specific heat of air, $[e_s\,(T_a) - e]$ is the difference in water vapour pressure between the ambient air (e) and the air at saturation $[e_s\,(T_a)]$, s is the rate of change of saturation vapour pressure with temperature, which is non-linear, and γ is the psychrometric constant (Monteith, 1973). The total of all the stomatal and boundary layer resistances of the individual leaves of the canopy are r_s and r_a, respectively. The boundary layer resistance r_a is the boundary layer resistance to water vapour, and r_{aH} to sensible heat. The boundary layer resistances to heat and water vapour are very similar and are often equated.

A number of the variables in the equation may be readily determined from standard meteorological measurements. Measurements of the net radiant balance however are rarely taken, although solar radiation, which is easier to measure, is widely available. Monteith (1973) and Rosenberg (1974) have shown that R_N, the net radiant balance (including the soil flux G and where $R_N = R + G$), may be readily estimated from S, the incoming solar radiation. The following empirical relationship has been adopted for predicting R_N:

$$R_N = 0.84S - 94, \qquad \textbf{(16)}$$

where R_N and S are measured in units of Wm^{-2}. Measurements of the soil heat flux G are required in order to measure R, the net radiant balance of the canopy. These are rarely achieved, however the measurements of G published in Monteith (1976), for a range of different vegetation types, indicate that G is a close to constant fraction of R_N:

$$G = 0.033R_N, \text{ and so} \qquad \textbf{(17)}$$

$$R = 0.967R_N. \qquad \textbf{(18)}$$

Some meteorological stations only record hours of sunshine, with no measurements of solar radiation. However, for the year it has been found, from the data presented in Müller (1982) and for defined areas, that solar radiation is strongly correlated with hours of sunshine. Solar radiation has been predicted in this way when such strong correlations emerge.

The aim has been to solve the Penman–Monteith equation for hourly intervals of an average day for each month of the year. The solar irradiance at each hour of the day has been predicted from published daily integrals by the following:

$$S = S_m(\sin\phi \sin\delta) + (\cos\phi \cos\delta \cos h), \qquad \textbf{(19)}$$

where S_m is the maximum daily irradiance, determined as:

$$S_m = (\Sigma S\pi)/2N, \qquad \textbf{(20)}$$

where ΣS is the published daily integral of irradiance and N is the daylength. Daylength is estimated from:

$$N = 2h_s/15, \text{ where} \qquad \textbf{(21)}$$

$$\cos h_s = -(\sin\phi \sin\delta)/(\cos\phi \cos\delta), \qquad \textbf{(22)}$$

and h is the time measured as an hour angle, h_s is the half-day length measured as an hour angle, ϕ is the latitude and δ the solar declination (Gates, 1980; Woodward & Sheehy, 1983).

Air temperature, like irradiance, follows a sine-wave pattern of change throughout the day, with a maximum about 2 h after midday and a minimum at about dawn. The typical daily range of temperature is calculated from the difference between the mean daily maximum and minimum temperatures of each month and these are adjusted to occur at dawn and at 2 h after midday, with a sine-wave change in temperature adjusted to conform to these times.

The diurnal pattern of absolute humidity does not appear to follow a regular pattern like irradiance and temperature, although a pattern which is inversely related to temperature is true for relative humidity. Measurements of relative humidity have been assumed to have been taken at 9.00 a.m. The temperature is predicted for this time and so the vapour pressure or absolute humidity may be obtained from the relationship between temperature and saturation vapour pressure, or absolute humidity. Rosenberg (1974) and Woodward & Sheehy (1983) provide equations which describe this curvilinear relationship. It has also been assumed that the absolute humidity, or dew point temperature, of the air remains the same throughout the day, with the vapour pressure deficit {$[e_s(T_a) - e)]$ in (**15**)} being determined by changes in air temperature.

The two least tractable features of the Penman–Monteith equation are the predictions of the canopy boundary layer and stomatal resistances. For a canopy with n layers of leaves, each of which will be transpiring, these resistances will all be in parallel; therefore, where $n = 3$ for example:

$$\frac{1}{r_s} = \frac{1}{r_1} + \frac{1}{r_2} + \frac{1}{r_3}, \qquad \textbf{(23)}$$

where r_s is the canopy stomatal resistance and $r_1 - r_3$ are the stomatal resistances of the leaves in each of the three layers.

More generally, if the leaf area index (L) is a measure of the number of leaf layers then the canopy resistance is a function of the stomatal resistance in each leaf layer (i):

$$\frac{1}{r_s} = \sum_{i=1}^{i=L} \frac{1}{r_i}. \qquad \textbf{(24)}$$

The aim of this model is to predict the maximum leaf area index which can develop at a site, given that the only limitation to leaf expansion is the availability of water, or in effect, drought. Grace, Okali & Fasehun (1982) have shown that the model works efficiently in reverse, when predicting the evapotranspiration of forest trees varying in leaf area index. It is clear from their measurements that the stomatal resistance of individual leaves will vary with depth in the canopy, with the lowest resistances at the canopy surface and the highest resistances at the bottom of the canopy. The gradient of vapour pressure between the leaf and air will decline with depth in the canopy, because of a reduction in radiant heating, air temperature and wind speed (McNaughton & Jarvis, 1983). Stomata respond strongly to the leaf to air vapour pressure deficit, with stomatal resistance increasing strongly between 0 and 5 KPa, for a wide range of species from differing physiognomic types of vegetation (Waring & Franklin, 1979; Schulze & Hall, 1982). So, in spite of an increased potential for transpiration in the greater vapour pressure deficits of the upper canopy, it is clear that other features of climate may control this response. Irradiance is also known to exert a strong effect in this way (e.g. Jarvis, 1976; Jones, 1983). The response of resistance to irradiance is hyperbolic, with a general form:

$$r = r_{min} + \frac{b}{S}, \qquad \textbf{(25)}$$

where r_{min} is the minimum stomatal resistance and b is a measure of the sensitivity of stomatal resistance to irradiance, S. The resistance r_{min} will be sensitive to the vapour pressure deficit (VPD) between the leaf and air but, in view of the uncertainties over predicting the vapour pressure of the air from meteorological records and of the precise nature of the VPD response of stomatal resistance, it has been decided to take a fixed minimum stomatal resistance of 400 sm^{-1}. This is on the basis of the review by Körner, Scheel & Bauer (1979), which shows a value of 400 sm^{-1} as a typical mean value, presumably itself dependent on VPD, for a range of vegetation types.

The sensitivity of stomatal resistance to irradiance (b in (**25**)) has a value of 29 500 sm^{-1} $(Wm^{-2})^{-1}$, which predicts significant increases in stomatal resistance at irradiances below about 200 Wm^{-2}.

Within a uniform plant canopy, the mean solar radiation may be calculated from the relationship:

$$S_z = S_0 \exp\left(\frac{-KL_z}{\sin \mathrm{a}}\right), \qquad (\mathbf{26})$$

where S_0 is the irradiance at the top of the canopy and Sz the irradiance at a distance z beneath the canopy surface and under an accumulated leaf area index Lz (Gates, 1980). K is the extinction coefficient for solar radiation and a is the solar elevation, where:

$$\sin a = (\sin \phi \sin \delta) + (\cos \phi \cos \delta \cos h). \qquad (\mathbf{27})$$

Therefore, if the disposition of leaf area index (L) with height and the mean canopy extinction coefficients are known, it is possible to predict both the solar radiation within the canopy at any hour of the day and its effect on stomatal resistance

Figure 4.3 represents the characteristic distribution of L and the penetration of irradiance for a range of L from 1 to 9, from a range of published information (Nomoto, 1964; Miller, 1966; Monsi, 1968; Kira, Shinozaki & Hozumi, 1969; Satoo, 1970; Kira, 1975; Cernusca, 1976; Monteith, 1976; Ashton, 1978; Dennis, Tieszen & Vetter, 1978; Kira, 1978; Aber, 1979; Walter, 1979; Houssard, Escarre & Romane, 1980; Vareschi, 1980).

The total leaf area has been divided into six strata: the lowest stratum (number 1) represents the herbaceous component of the canopy, with strata 2 and 3 the boundary beneath the main canopy crown, which has its greatest value of L in stratum 5.

The profiles of irradiance are values from a range of field observations. The irradiance profiles include an allowance for canopy reflectivity, which has a mean value of 16%. The extinction coefficient has a value of 0.5 at a leaf area index of 5, decreasing to 0.48 at a leaf area index of 7 and 0.45 at a leaf area index of 9. As Jarvis & Leverenz (1983) show, there is some degree of variation in the extinction coefficient between the canopies of different species, although the mean value of their measurements is 0.47 and close to those above. The chosen values are therefore simplifications of reality but can readily be modified in the light of more extensive information. Given the measurements of L and irradiance with depth in the canopy and at different times of day, it is then possible, using (**26**), to predict the stomatal resistance in any stratum of the canopy, assuming

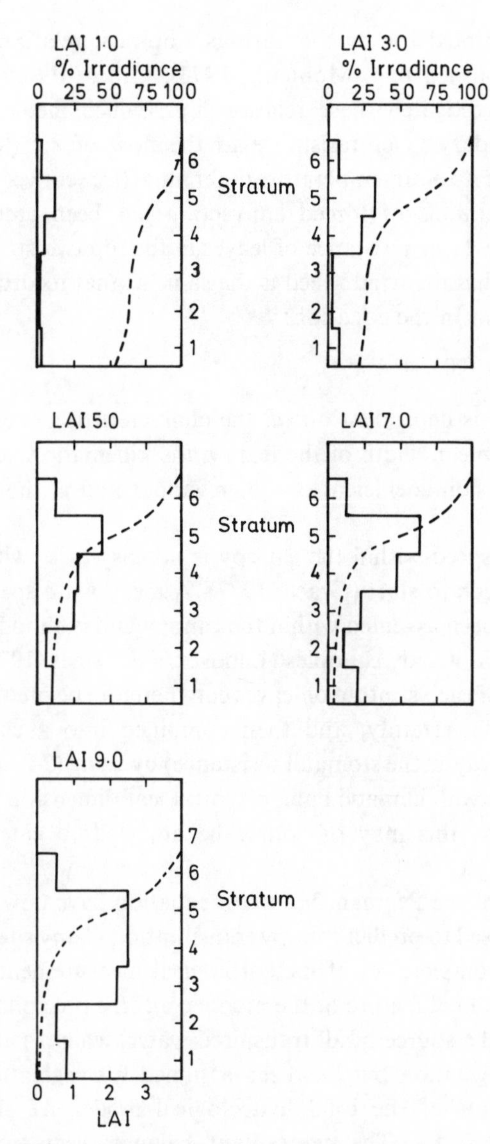

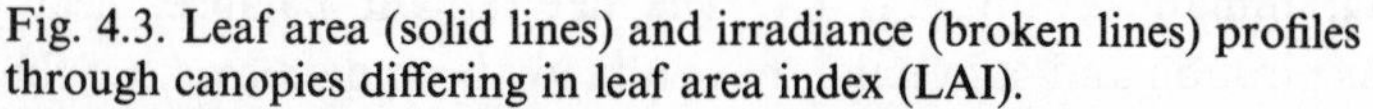
Fig. 4.3. Leaf area (solid lines) and irradiance (broken lines) profiles through canopies differing in leaf area index (LAI).

control only by irradiance (**25**). The canopy stomatal resistance may then be determined using (**24**).

The final component of the Penman–Monteith equation which must be predicted is the canopy boundary layer resistance. This resistance is dependent on both leaf size and wind speed. The effect of these two variables may be accounted for by the various empirical relationships which have been established (e.g. Monteith, 1973). Grace, Fasehun & Dixon (1980) have investigated these relationships, which include the responses of the boundary layer resistance to the flow of air (forced convection) and to the leaf-to-air temperature differential (free convection). The equation for conditions of forced convection has been used for predicting the boundary layer resistance of leaves in the top two strata of the canopy, assuming that the wind speed is the same as that recorded at a meteorological station. In the equation:

$$r_a = d^{0.2}\, \nu^{0.25} / (0.03\, D^{0.67} u^{0.8}), \qquad \textbf{(28)}$$

the boundary layer, r_a, is dependent on: d, the characteristic dimension which is related to the mean width of the leaf; ν, the kinematic viscosity of dry air; D, the diffusion coefficient of water vapour and u, the wind speed.

Predicting the wind speed within the canopy is no easy task, with no 'general case' from which to start (Grace, 1977). A mean wind speed of 0.4 ms^{-1} has therefore been assumed within the canopy and is a moderate match to observations in woody canopies (Landsberg & James, 1971).

The boundary layer of each stratum of leaves can therefore be predicted, with some degree of uncertainty, and then combined into a canopy resistance (in the same way as the stomatal resistance) by using (**24**). When the model has been run with climatic data, a typical leaf dimension of 50 mm has been assumed; this may of course be adjusted to cater for needle-leaves, for example.

All the components of the Penman–Monteith equation have now been described and may be used to predict canopy transpiration at any site over the globe at which a complete set of meteorological measurements are recorded regularly. One final feature of the model concerns precipitation. This is assumed to be the source of all transpired water, which is clearly in error for riverine vegetation but has been assumed for vegetation in general. The components of the total hydrological model are shown diagrammatically in Fig. 4.4. The net radiant balance, heat storage, transpiration and convection have already been discussed. Of the rainfall that falls on to the canopy, some penetrates to the soil as throughfall, whilst some is intercepted and remains as a film of surface water on leaves, branches and trunks. Of the intercepted component, some water evaporates whilst the remainder is considered to contribute to throughfall.

Evaporation of the water film from leaves may account for a considerable proportion of the intercepted water because the only resistance to evaporation is the boundary layer resistance. This high rate of evaporation

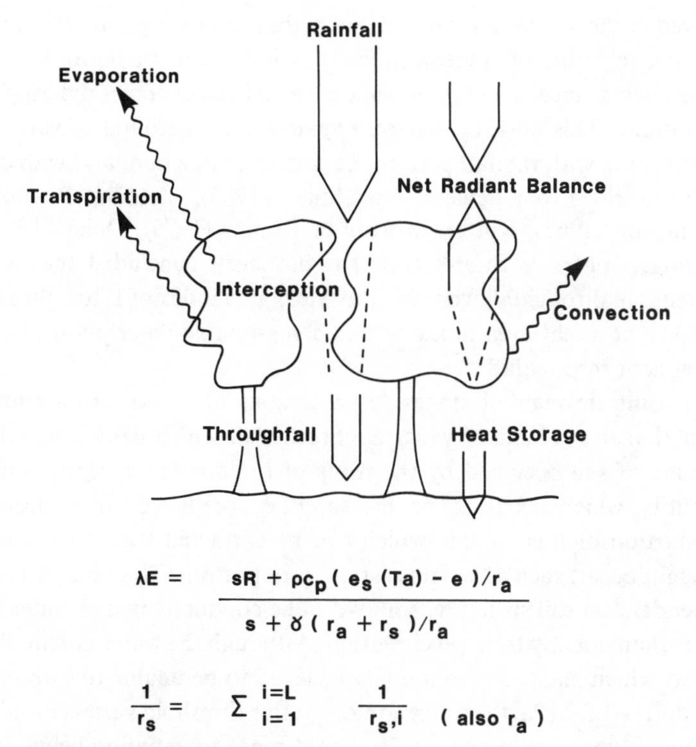

Fig. 4.4. Schematic of the components of the hydrological and energy balance of a forest canopy.

continues until the film evaporates or drips to the soil beneath. In this respect, the drip tips characteristic of many tropical leaves are considered to enhance canopy throughfall (Dean & Smith, 1979).

The meteorological records of precipitation are presented as monthly totals, with no indication of any characteristic diurnal periodicity in rainfall. For the purposes of the predictive model, therefore, the monthly figures have been converted to hourly figures, with evaporation calculated for each hour of the day, using the Penman–Monteith equation and a stomatal resistance set to zero. Evaporation proceeds in this manner until the integrated evaporation equals interception, i.e. all of the water film has just evaporated, from which point transpiration through stomata is the rule.

Interception or throughfall must in some way be related to leaf area index, with a decrease in throughfall with increasing leaf area index the most likely relationship. The proportion of precipitation as throughfall is also critical because the throughfall of water to the soil is assumed to be the only source of water for transpiration and must in some way be used to predict the maximum leaf area index that can be supported by the local climate, in terms of transpiration. It is important therefore to provide some average measure of interception losses from canopies differing in leaf area index. This must be achieved against a background of variation in interception with rainfall pattern, i.e. storm or fine rain, and with canopy architecture. From reviews by Zinke (1967), Leyton, Reynolds & Thompson (1967), Rutter, Morton & Robins (1975), Doley (1981) and Waring, Rogers & Swank (1981) it has been concluded that a linear decrease in throughfall from 95% at a leaf area index of 1, to a throughfall of 83% at a leaf area index of 9 is a reasonable description of canopy dependent throughfall.

In using throughfall alongside predictions of evapotranspiration, it is clear that the volume of water as throughfall which percolates into that volume of soil occupied by the roots of the dominant vegetation is the quantity which needs to be measured or predicted. In addition, the proportion of this volume which can be extracted before the effects of drought occur, such as leaf abscission, is important. This volume is closely dependent on soil structure, soil hydraulic conductivity and water supply other than directly from precipitation. Although the water potential of the soil at which plants are commonly believed to be unable to extract water is about −1.5 MPa, the water content at this threshold varies considerably with soil type, with perhaps a 2 – 3-fold range in variation being possible (Miller, 1977; Waring *et al.*, 1981). The ideal hydrological model should therefore include edaphic features to account for this. This has not been attempted because of the paucity of global information. Penman (1963) has assumed that plants transpire all water from the rooting zone, plus an additional 25 mm of water (rainfall equivalent measurement of volume) drawn from below this level. Priestley & Taylor (1972) and Lockwood (1979) have measured a rainfall equivalent of 75 mm of water available for transpiration. This total has been rounded to 80 mm and is used as a threshold for drought in the model. Therefore when transpiration exceeds throughfall, such that the time integral of the hydrological budget reaches a deficit of 80 mm of water, then leaf area reduction by abscission should occur.

The implementation of the model is shown graphically on Fig. 4.5. The upper diagram shows the computed monthly soil water deficit at Brisbane, Australia for odd numbered leaf area indices (L) between 1 and 9. The computations start in January, at which time precipitation exceeds evapotranspiration for all leaf area indices and so no soil water deficit occurs and run-off is expected to be the dominant hydrological process. This state of affairs occurs through the year for a leaf area index of 1. Deficits are

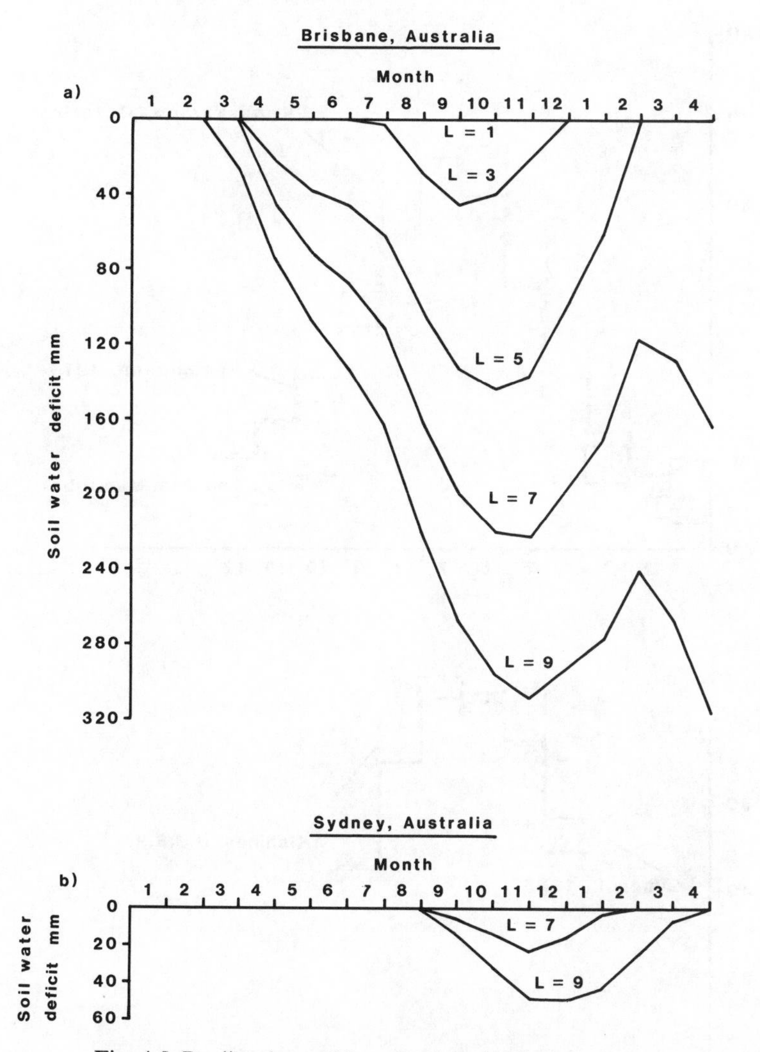

Fig. 4.5. Predicted monthly soil water deficit for soils at two sites in Australia: (*a*) Brisbane; (*b*) Sydney.

however predicted to occur from the end of February ($L = 9$) and March ($L = 5$ and 7). No significant deficit is computed until the end of July for a leaf area index of 3. When the predictions are carried on to the end of March of the second year then it may be seen that there is no recharge of soil water under leaf area indices of 7 and 9, with a continual drying being expected. Soil recharge is predicted for leaf area indices of 5 and less. The fact that the soil water deficit exceeds 80 mm of rainfall equivalent for a leaf area index of 5 suggests that the leaf area index for this site

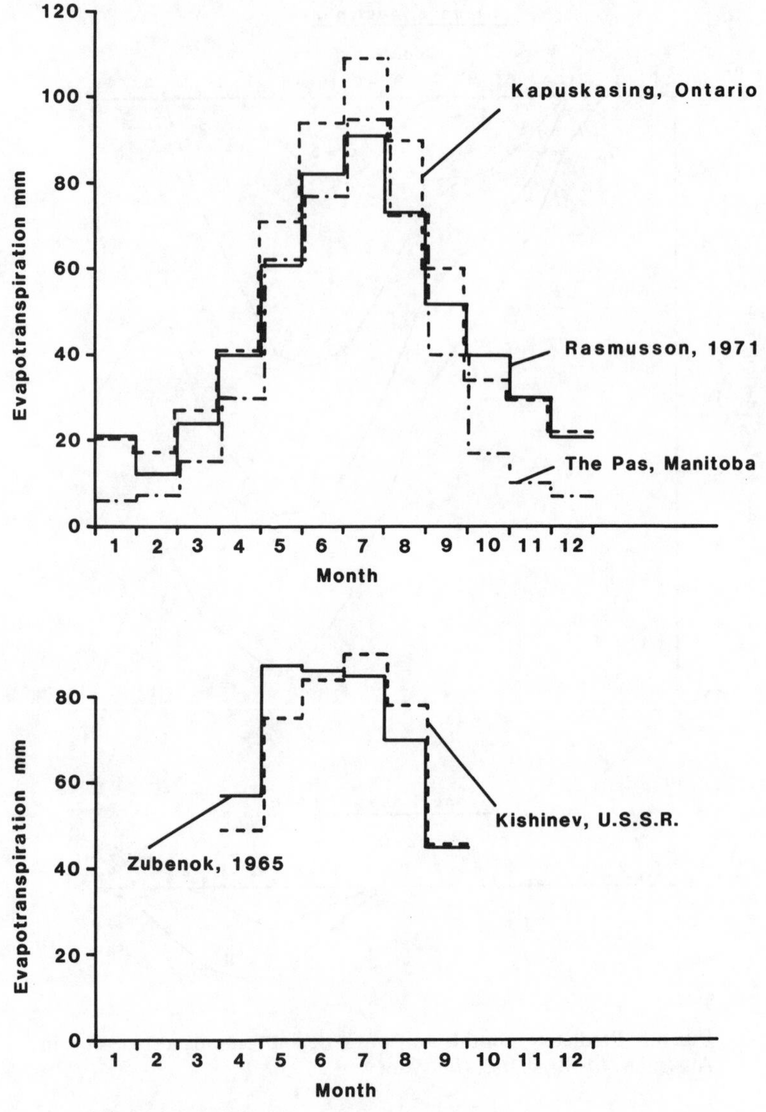

Fig. 4.6. Model test.

should be between 3 (maximum soil water deficit of 45 mm) and 5. For Sydney, shown in Fig. 4.5(*b*), the November and December droughts are less severe and so a higher leaf area index of 9 is predicted.

Tests of the hydrological model

A test of the predictive efficiency of the hydrological model is shown on Fig. 4.6. Rasmusson (1971) has determined the average upward

Table 4.2 *Evapotranspiration comparisons for Russian stations*

Site	Latitude	Evapotranspiration (mm year^{-1})	
		Prediction	Rauner computation
Turgaj	45 °38′ N	296	300
Moscow	55 °45′ N	530	550
Archangel'sk	64 °30′ N	381	370
Dudinka	69 °24′ N	180	240

transfer of water vapour from the forest vegetation of eastern North America. These observations are shown on Fig. 4.6(*a*), alongside predictions from the centre of this range for Kapuskasing, Ontario, Canada, with a predicted leaf area index of 9, and for The Pas, Manitoba, Canada, with a predicted leaf area index of 7. Predictions further north have lower annual totals for evapotranspiration and predictions from further south have greater annual totals, resulting particularly from differences in the growing season; however it is clear that there is a close match between observation and prediction.

Zubenok (1974) has measured evapotranspiration from grassland in the Ukraine, USSR at L'vov. These data have been compared with predictions at a similar site at Kishinev (USSR) with a predicted leaf area index of 3. It is again clear that measurement and prediction correspond quite closely.

Rauner (1972, and in Miller, 1977) has measured and computed average annual totals of evapotranspiration from vegetation in western Russia. Annual totals extrapolated from a map of these measurements are compared in Table 4.2 with predicted values from four stations, in the centre of these vegetational zones. Close correspondence is again notable for vegetation types ranging from tundra, through boreal forest to grassland or steppe. Similar correspondence can also be observed with the records of Budyko (1974).

Global predictions

These few tests of the model suggest that it is at least adequate in predicting evapotranspiration from standard meteorological records. These records have therefore been used to predict the leaf area index for meteorological stations over the world's surface, using the criteria described earlier. The results of these predictions are presented on the first map, Fig. 4.7. The size of the circles is related to the leaf area index, to an upper limit

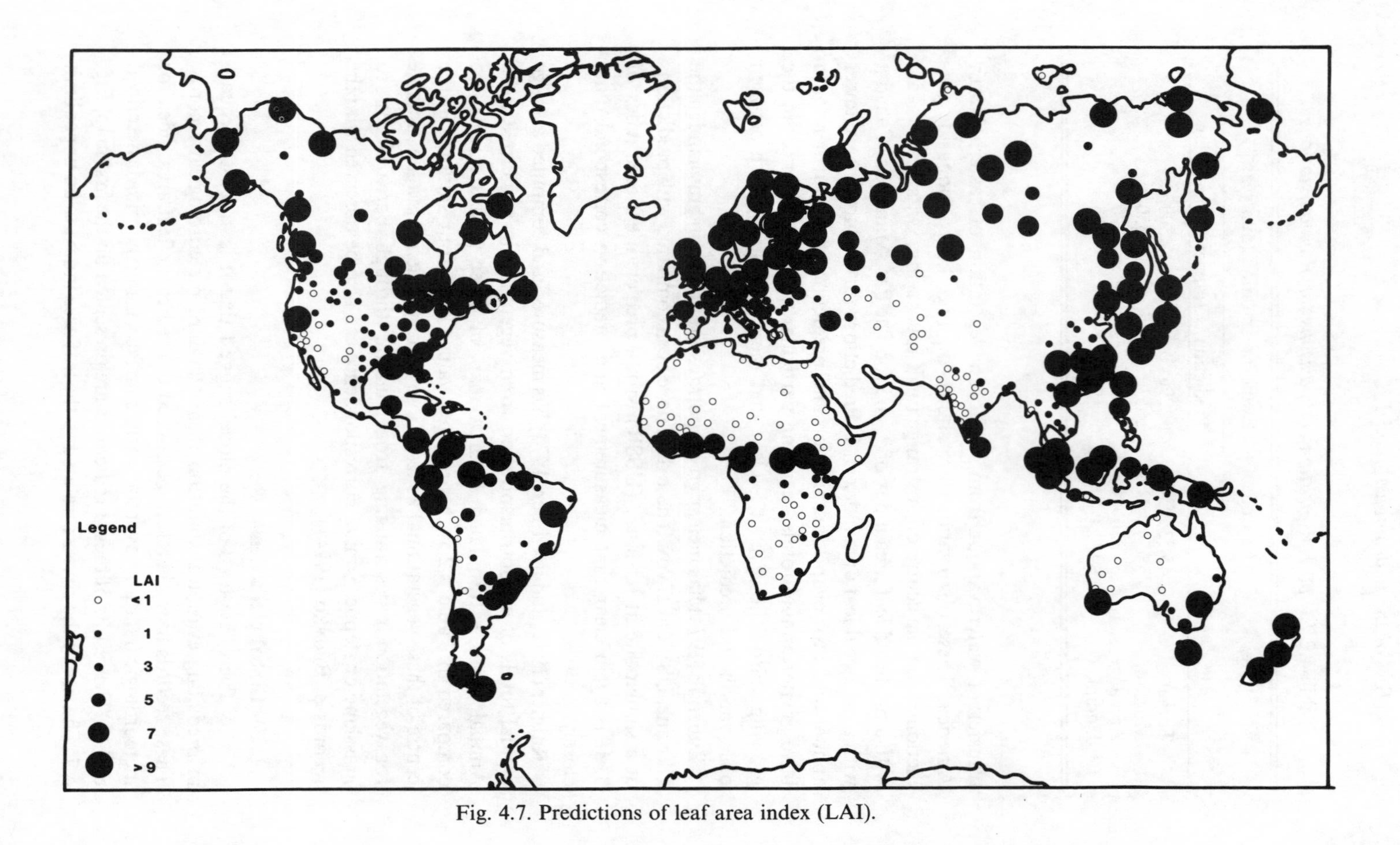

Fig. 4.7. Predictions of leaf area index (LAI).

of 9 or greater. The open circles represent predictions of a leaf area index of less than 1.

The geographical distribution of vegetation based on these predictions may be compared with a vegetation map of the world. A reduced version of the detailed map prepared by A. W. Küchler for Polunin (1960) has been used for comparison.

The predictions of leaf area index are closely correlated with observed physiognomic types of vegetation. High leaf area indices correspond with the equatorial rain forests of central Africa, South America and Indonesia; the tropical forests of South America and southeast Asia; the cool temperate forests of North America, Europe, Russia, eastern Asia, South America and Australasia, and finally with the cold boreal zones of North America and Russia.

The prediction of a range in leaf area index between 1 and 3 corresponds with savannah vegetation (Africa, South America, southeast Asia) and with grassland (the prairies or steppes of North and South America and the USSR). Predictions of a leaf area index less than 1 correspond with the deserts of central North America, western South America, central Asia, Africa and Australia.

These predictions of leaf area index are totally dependent on the local balance between precipitation and evapotranspiration. Low rates of evapotranspiration will occur in areas with very high ambient humidity, which may be realised in both cold and warm regions. Precipitation on the other hand generally decreases with latitude. However, it is clearly possible to predict high leaf area indices at high latitudes because of the very low rates of evapotranspiration, which rarely exceed precipitation. The model predicts high leaf area indices along the northern coasts of North America and Russia, although these are areas of tundra with a characteristically low leaf area index (Tieszen, 1978). It appears that the limited and low-temperature growing season is more critical to leaf development than the hydrological budget. Other areas where the model predictions are at variance with vegetation stature are in regions of strongly seasonal rainfall, such as in northern India, southeast Asia and central Africa. Here, leaf area indices are higher than predicted. In these regions there is a significant dry season of some months, during which time the predicted soil water deficit is high (as in the examples on Fig. 4.5(*a*)). The wet season may be strongly contrasting or monsoon-like, with abundant rainfall to fund evapotranspiration by forests with high leaf area indices. It is predicted for these areas that the vegetation should be predominantly broad-leaved and drought deciduous, with leaf development occurring rapidly in the warm temperatures experienced close to the rainy seasons.

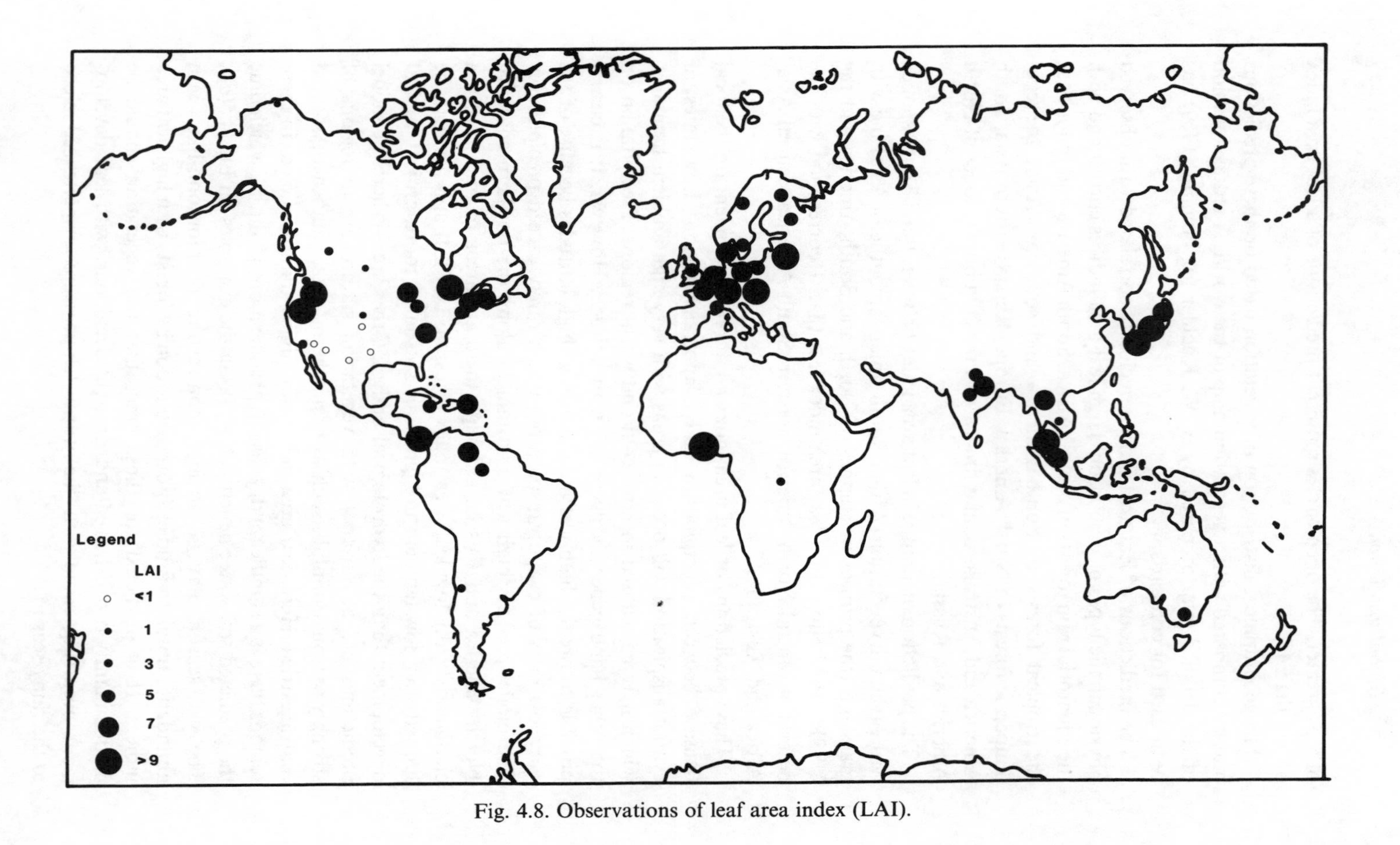

Fig. 4.8. Observations of leaf area index (LAI).

A further test of the predictions of leaf area index is provided on the second map, Fig. 4.8, which displays the rather limited range of actual observations on leaf area index in the world's natural vegetation. These observations have been taken from Cannell (1982) and from references in Schulze (1982).

In spite of the poor coverage, the measurements serve to emphasise the earlier points of high leaf area indices in the humid equatorial, tropical and temperate regions and low leaf area indices in the drier African savannah and North American prairies. The observations also emphasise the errors of prediction in the tundra zone, with low leaf area indices in Alaska and Greenland, and in the areas of seasonal precipitation such as southeast Asia where leaf area indices are higher than predicted.

The errors in the simple predictions of leaf area index can be included in a further development of the model, which should also include the use of minimum temperatures to discriminate between physiognomic types of vegetation.

At latitudes which in general are beyond the Arctic Circle it is likely that the short growing season will be a severe limitation to the growth of leaves, and therefore establishment of leaf area index. This will be also true for shrubs, which are protected from very low temperatures by the presence of a deep snow cover. Snow cover may provide an ameliorated thermal protection to the extent of a 30 °C temperature differential between air and ground surface temperature (Ylimaki, 1962). Minimum temperatures at ground level may not necessarily exceed the −15 °C threshold for broad-leaved evergreen species. It is possible, therefore, to explain the presence of both broad-leaved evergreen and deciduous shrubs in tundra as dependent on the temperature differential between the air and the ground surface and the extent of the snow cover. However, the snow cover and temperature differential will be strongly dependent on local topography and can not be predicted from climatic records. Tundra, then, is assumed to be composed of a mixture of broad-leaved evergreen and deciduous species.

The heat sum, which is the integral of temperature and time required for the leaves of species from the tundra to expand, is readily comprehensible in physiological terms and can be computed from meteorological records. Observations on leaf expansion by a number of evergreen and deciduous shrubs from the tundra, e.g. *Salix brachycarpa*, *S. glauca*, *S. planifolia*, *Dryas integrifolia* and *Betula nana* (Bliss, 1956; Callaghan & Collins, 1981; Tieszen *et al.*, 1981), have been combined with measurements of air temperature to determine the number of day-degrees (mean daily temperature times number of days) for the completion of leaf expansion.

The sum of day-degrees is based on a threshold temperature of 0 °C. The mean temperature sum for these shrubs is 220 day-degrees, with values less than 200 for herbaceous species (Callaghan, 1974).

The greatest poleward extent of forest is to a latitude of 72 °50′N for the deciduous conifer *Larix dahurica* (Webber & Klein, 1977). It is obviously important, when considering leaf expansion of tundra shrubs, to determine also the difference between shrubs of the tundra and trees from the boreal zone in order to establish the basis of any quantitative differences. These differences will be critical in determining the northern spread of the trees of coniferous forest; trees which can endure the winter cold and desiccation (Hadley & Smith, 1983) may not be able to complete leaf expansion. Estimates of the heat sums for leaf expansion by the deciduous conifers *Larix lyallii*, *L. dahurica*, *L. decidua* and *L. sibirica* have been approximated from data of Arno & Habeck (1972), Webber & Klein (1977), Odum (1979), Tranquillini (1979), and Benecke & Havranek (1980). The mean heat sum is 610 day-degrees, substantially greater than the shrubs of the tundra, suggesting that the heat sum for the growing season may be critical in limiting the poleward spread of boreal forest.

The heat sum for leaf expansion of the evergreen conifers *Picea abies*, *P. mariana*, *Pinus contorta*, *P. mugo* and *P. sylvestris* have been estimated from the data of Gaertner (1964), Kozlowski (1964), Benecke & Havranek (1980), Black & Bliss (1980) and Beadle, Talbot & Jarvis (1982). The heat sum contrasts strikingly with the deciduous conifers with a mean of 950 day-degrees.

These three sums have been included into the predictive model in order to define the poleward extent of boreal forest and tundra. Tundra is predicted in areas where the heat sum is less than 600 day-degrees, deciduous conifers are predicted for sites where the heat sum is 600 to about 950 day-degrees, whilst the evergreen conifers are predicted where the heat sum is greater than 950 day-degrees. These predictions serve as a general test of the use of heat sums for predicting leaf development. All the possible heat sums for native species have not been covered and so any species with, for example, a heat sum for leaf expansion of 610 day-degrees could complete leaf expansion where a species of *Larix* is predicted.

Given more detail on leaf development of the tundra, it might also be possible to subdivide the tundra into shrub, herbaceous and thallophytic zones, but this has not been attempted.

The distinction between the vegetation types of the cold regions on the basis of heat sums, the delimitation of drought deciduous zones on the basis of rainfall and seasonality, and the definition of winter deciduous vegetation in terms of the winter minimum temperature have all been included in the model. In addition, the timing of the onset of leaf expansion in the spring and leaf abscission in the autumn has been predicted from the time when the minimum temperature rises or falls, respectively, below −5 °C. This threshold temperature has been established by comparing phenological records for deciduous trees (Flint, 1974; Lieth, 1974; Nienstaedt, 1974; Taylor, 1974; Schnelle, Baumgartner & Freitag, 1984) with local records of climate. These predictions are shown in the third map, Fig. 4.9.

The correspondences between prediction and observation are now much closer, emphasising the drought deciduous regions of Africa, India and America, the winter deciduous forests of Europe, western Russia and eastern Northern America and the deciduous coniferous forests of north Russia, although Polunin (1960) shows these to be more extensive in eastern Russia.

The major errors in prediction fall in North America, and in particular by predicting a winter deciduous forest along the Pacific northwest where an evergreen coniferous forest is dominant. Similar errors in prediction are also true for the maritime areas of Scandinavia and the British Isles where more southerly vegetation types are predicted. In additon, the prairies of North America have been subdivided into winter-deciduous in the south and evergreen coniferous in the north. This is because of a distinct lack of information on the winter threshold temperatures for survival of grasses which may be covered by snow. However, forests have a low leaf area index threshold of about 3 (Jarvis & Leverenz, 1983) so that shrubs or a mixture of shrubs and grasses are predicted for these areas. The model, however, has no fundamental basis for predicting the presence of shrubs or of herbaceous species.

Waring & Franklin (1979) have discussed the reasons for the dominance of the Pacific northwest by evergreen, but temperate, coniferous forests. They consider that such a forest has established itself because the forests escaped severe mortality during the glaciations of the Pleistocene. In addition the rather mild winters are followed by warm dry summers which would lead to drought and reduce the potential for growth by broad-leaved deciduous species. The evergreen conifers can endure these periods of drought and are capable of growth in the cooler, damper months. In addition it is interesting to note that many of these species of evergreen conifer, e.g. *Abies grandis*, *Picea sitchensis*, *Pseudotsuga menziesii* and *Tsuga heterophylla* have leaf- and bud-freezing resistances of no less than −35 °C (Sakai & Weiser, 1973), falling firmly in the range of winter tolerance of broad-leaved deciduous species. In some respects the model predictions are correct in predicting the presence of a forest with a freezing

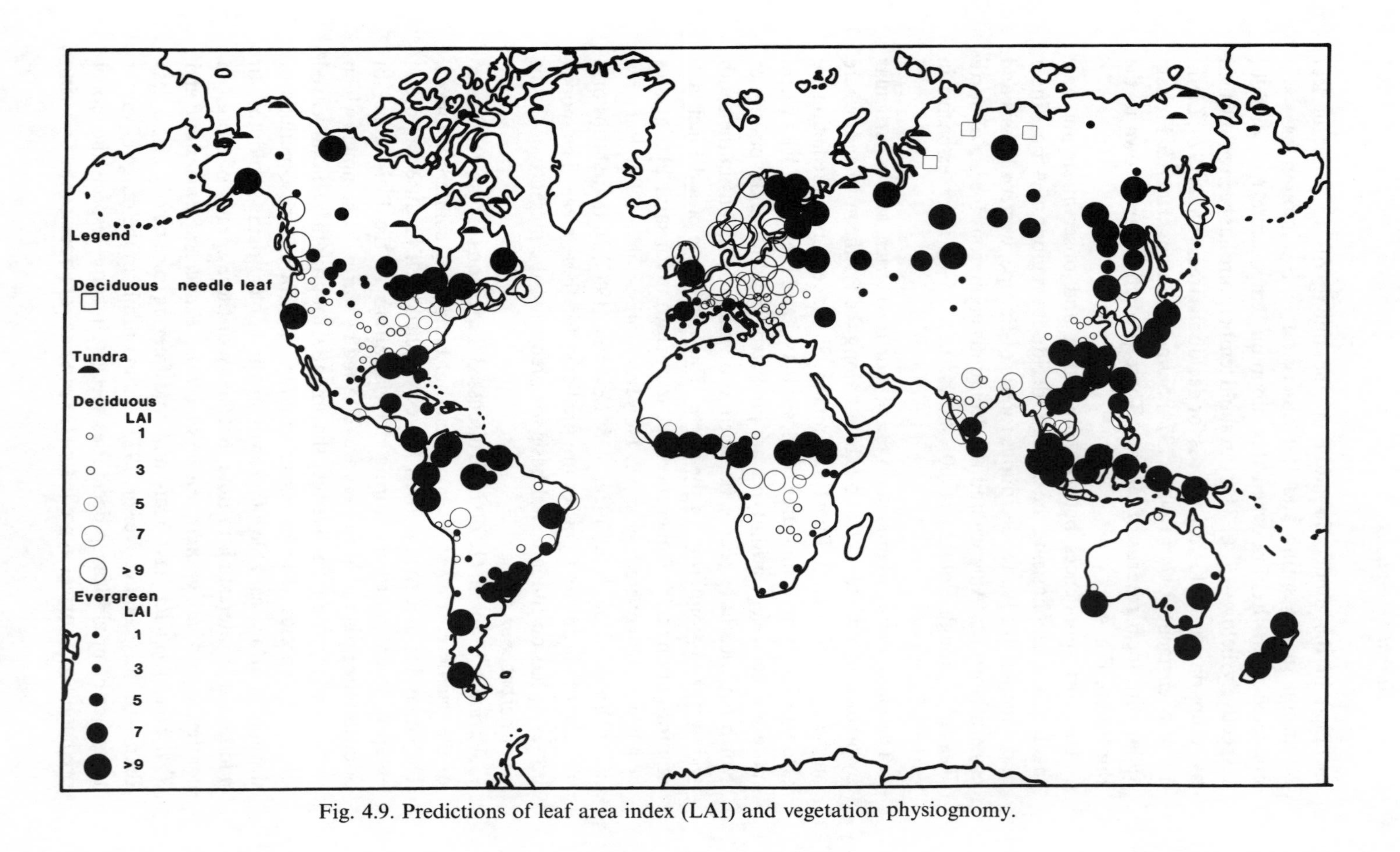

Fig. 4.9. Predictions of leaf area index (LAI) and vegetation physiognomy.

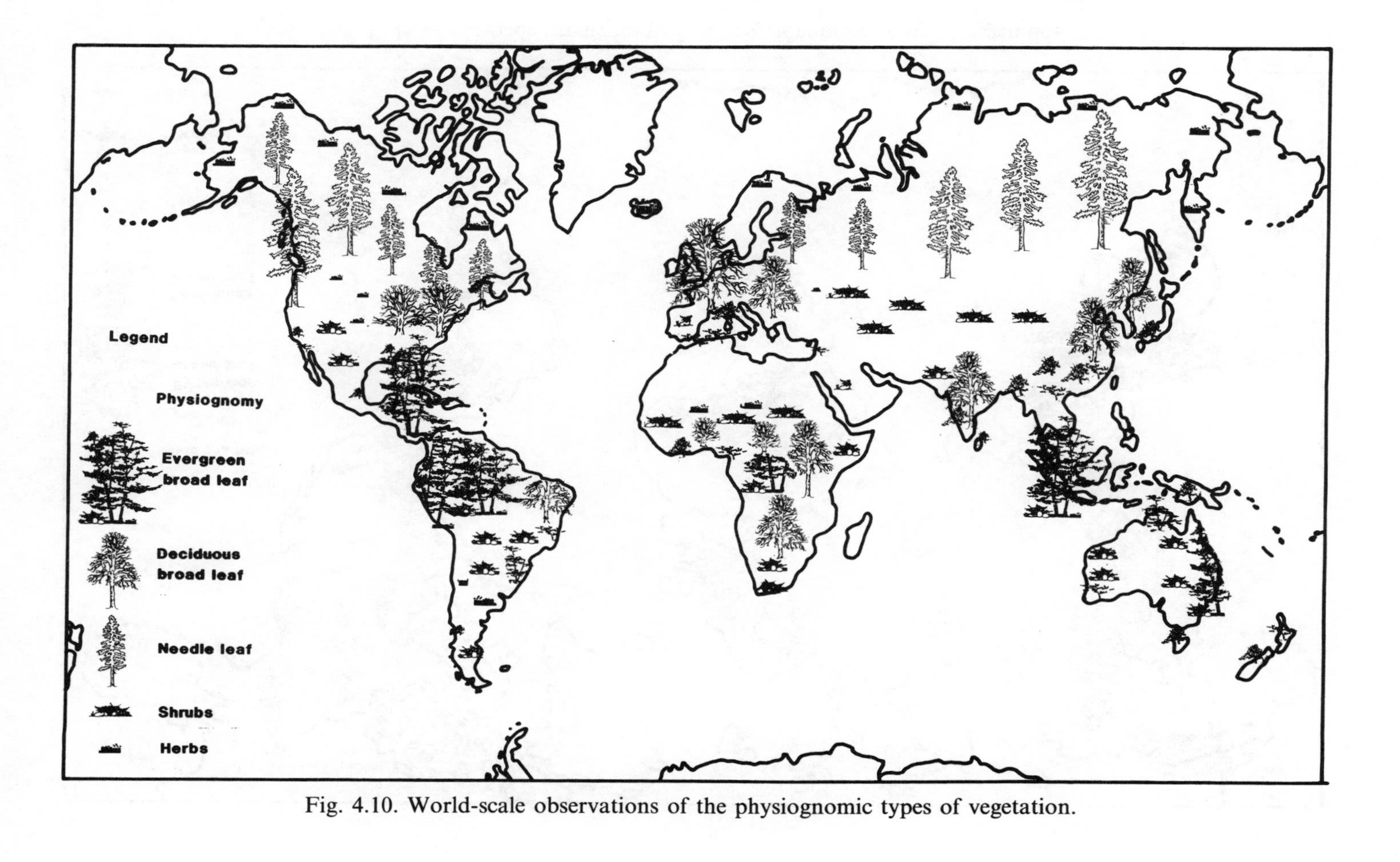

Fig. 4.10. World-scale observations of the physiognomic types of vegetation.

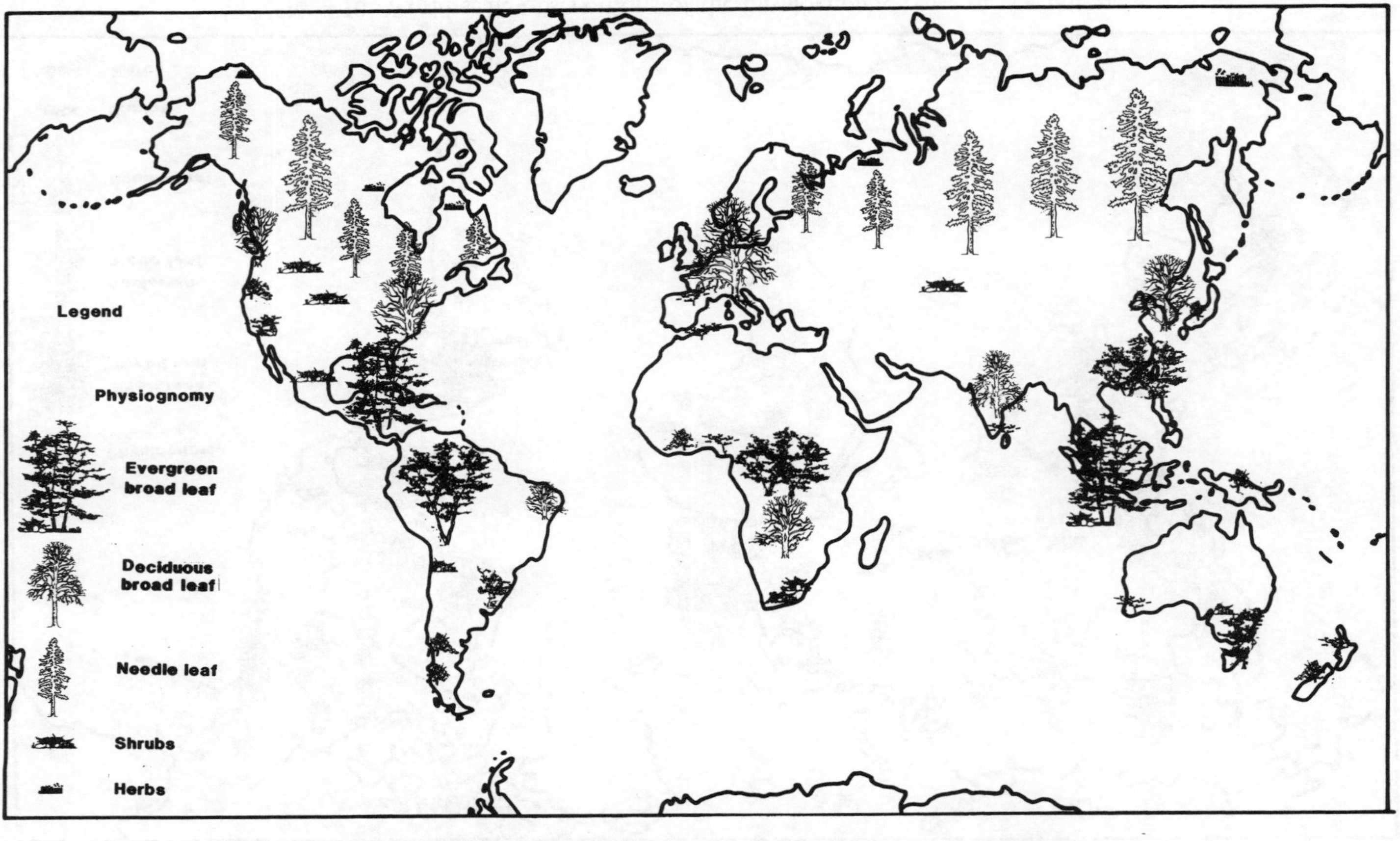

Fig. 4.11. World-scale predictions of the physiognomic types of vegetation.

resistance to about −40 °C but, as outlined earlier, certain combinations of climate and plant physiology may favour an unexpected competitive result.

The broad, world-scale patterns of the major physiognomic types of vegetation have been presented on Fig. 4.10 and have been taken, in much reduced form, from Polunin (1960). The predicted types are shown in Fig. 4.11; herbaceous vegetation is arbitrarily assumed to have a leaf area index of 1, with shrubs having a leaf area index of 3 and forests a leaf area index of greater than 3. Tundra embraces a large range of physiognomy and is described 'on average' as herbaceous on both maps.

The maps show that the model predictions are, with the exceptions mentioned above, extraordinarily accurate, yet they are based on physiological principles. Mechanisms for the control of plant distribution have been presented but their validation, by experiment and observation, is now required. The problem here is that the studies should be on vegetation, not on individual species, and vegetation is an unwieldy subject for experimentation.

References

Aber, J.D. (1979) Foliage-height profiles and succession in northern hardwood forests. *Ecology*, **60**, 18–23.

Addicott, F.T. & Lyons, J.L. (1973). Physiological ecology of abscission. In *Shedding of Plant Parts*, ed. T.T. Kozlowski, pp. 85–124. New York: Academic Press.

Arno, S.F. & Habeck, J.R. (1972). Ecology of alpine larch (*Larix lyallii* Parl.) in the Pacific northwest. *Ecological Monographs*, **42**, 417–50.

Ashton, P.S. (1978). Crown characteristics of tropical trees. In *Tropical Trees as Living Systems*. ed. P.B. Tomlinson & H.H. Zimmermann, pp. 591–615. Cambridge University Press.

Auld, B.A., Dennett, M.D. & Elston, J. (1978). The effect of temperature changes in the expansion of individual leaves of *Vicia faba*: 1. *Annals of Botany*, **42**, 877–88.

Axelrod, D.I. (1966). Origin of deciduous and evergreen habits in temperate forests. *Evolution*, **20**, 1–15.

Bate, G.C., Furniss, P.R. & Pendle, B.G. (1982). Water relations of southern African savannahs. In *Ecology of Tropical Savannahs*. ed. B.J. Huntley & B. H. Walker, pp. 336–58. Berlin: Springer-Verlag.

Beadle, C.L., Talbot, H. & Jarvis, P.G. (1982). Canopy structure and leaf area index in a mature Scots pine forest. *Forestry*, **55**, 105–23.

Becwar, M.R. & Burke, M.J. (1982). Winter hardiness limitations and physiography of woody timberline flora. In *Plant Cold Hardiness and Freezing Stress: Mechanisms and Crop Implications*, vol.2, ed. P.H. Li & A. Sakai, pp. 307–23. New York: Academic Press.

Benecke, U. & Havranek, W. M. (1980). Gas exchange of trees at altitudes up to timberline, Craigieburn Range, New Zealand. In *Mountain Environments and Subalpine Tree Growth*, ed. U. Benecke & M.R. Davis, pp. 195–212. New Zealand Forest Service: Forest Research Institute.

Bennett, K.D. (1984). The post glacial history of *Pinus sylvestris* in the British Isles. *Quaternary Science Reviews*, **3**, 133–55.

Black, R.A. & Bliss, L.C. (1980). Reproductive ecology of *Picea mariana* (Mill.) BsPl., at tree line near Inuvik, Northwest Territories, Canada. *Ecological Monographs*, **50**, 331–54.

Bliss, L.C. (1956). A comparison of plant development in microenvironments of arctic and alpine tundra. *Ecological Monographs*, **26**, 303–37.

Botkin, D.B., Janak, J.F. & Wallis, J.R. (1972). Some ecological consequences of a computer model of forest growth. *Journal of Ecology*, **60**, 849–72.

Box, E.O. (1981). *Macroclimate and Plant Forms: An Introduction to Predictive Modeling in Phytogeography*. The Hague: Junk.

Boyer, J.S. (1968). Relationship of water potential to growth of leaves. *Plant Physiology*, **43**, 1056–62.

Budyko, M.I. (1974). *Climate and Life*. New York: Academic Press.

Burke, M.J., Gusta, L.V., Quamme, H.A., Weiser, C.J. & Li, P.H. (1976). Freezing injury in plants. *Annual Review of Plant Physiology*, **27**, 507–28.

Bussell, W.T. (1968*a*). The growth of some New Zealand trees. 1. Growth in natural conditions. *New Zealand Journal of Botany*, **6**, 63–75.

Bussell, W.T. (1968b). The growth of some New Zealand trees. 2. Effects of photoperiod and temperature. *New Zealand Journal of Botany*. **6**, 76–85.

Cain, S.A. (1944). *Foundations of Plant Geography*. New York: Harper.

Callaghan, T.V. (1974). Intraspecific variation in *Phleum alpinum* L. with specific reference to polar populations. *Arctic and Alpine Research*, **6**, 361–401.

Callaghan, T.V. & Collins, N.J. (1981). Life cycles, population dynamics and the growth of tundra plants. In *Tundra Ecosystems: a Comparative Analysis*, ed. L.C. Bliss, O.W. Heal & J.J. Moore, pp. 257–84. Cambridge University Press.

Candolle, A.I. de (1855). *Géographique Botanique Raisonée*. Paris: Masson.

Cannell, M.G.R. (1982). *World Forest Biomass and Primary Production Data*. New York: Academic Press.

Cernusca, A. (1976). Bestandesstruktur, Bioklima und Energiehaushalt von alpinen Zergstrauchbeständen. *Oecologia Plantarum*, **11**, 71–102.

Chabot, B.F. & Hicks, D.J. (1982). The ecology of leaf lifespans. *Annual Review of Ecology and Systematics*, **13**, 229–59.

Cockayne, L. (1928). *The vegetation of New Zealand*, 2nd edn. Leipzig: Englemann.

Cosgrove, D. (1984). Hydraulic aspects of plant growth. *What's New In Plant Physiology*, **15**, 5–8.

Davis, M.B. (1980). Quaternary history and the stability of forest communities. In *Forest Succession Concepts and Application*, ed. D.C. West, H.H. Shugart & D.B. Botkin, pp. 132–53. New York: Springer-Verlag.

Dean, J.M. & Smith, A.P. (1979). Behavioral and morphological adaptation of a tropical plant to high rainfall. *Biotropica*, **10**, 152–4.

Dennis, J.G., Tieszen, L.L. & Vetter, M.A. (1978). Seasonal dynamics of above- and below-ground production of vascular plants at Barrow, Alaska. In *Vegetation and Production Ecology of an Alaskan Arctic Tundra*. ed. L.L. Tieszen, pp. 113–40. New York: Springer-Verlag.

Doley, D. (1981). Tropical and subtropical forests and woodlands. In *Water Deficits and Plant Growth*. vol. VI, ed. T.T. Kozlowski, pp. 209–323. New York: Academic Press.

Dumbelton, L.J. (1967). Winter dormancy in New Zealand biota and its paleoclimatic implications. *New Zealand Journal of Botany*, **5**, 211–22.

Ellenberg, H. (1978). *Vegetation Mitteleuropas mit den Alpen in ökologischer Sicht.* Stuttgart: Eugen Ulmer.

Ewers, F.W. & Schmid, R. (1981). Longevity of needle fascicles of *Pinus longaeva* (Bristlecone Pine) and other North American Pines. *Oecologia*, **51**, 107–15.

Flint, H.L. (1974). Phenology and genecology of woody plants. In *Phenology and Seasonality Modelling*, ed. H. Lieth, pp. 83–97. Berlin: Springer-Verlag.

Franks, F. (1982). Physiological water stress. In *Biophysics of Water*, ed. F. Franks & S. F. Mathias, pp. 279–94. Chichester: Wiley.

Franks, F. (1983). Cold stress and resistance in plants. *What's New in Plant Physiology*, **14**, 37–40.

Franks, F., Mathias, S.F. & Trafford, K. (1984). The nucleation of ice in undercooled water and aqueous polymer solutions. *Colloids and Surfaces*, **11**, 275–85.

Fukai, S. & Silsbury, J.H. (1976). Response of subterranean clover communities to temperature I. Dry matter products and plant morphogenesis. *Australian Journal of Plant Physiology*, **3**, 527–43

Gaertner, E.E. (1964). Tree growth in relation to the environment. *Botanical Review*, **30**, 335–92.

Gates, D.M. (1980). *Biophysical Ecology*. New York: Springer-Verlag.

Gates, D.M. & Papian, L.E. (1971). *Atlas of Energy Budgets of Plant Leaves*. New York: Academic Press.

George, M.F. (1982). Freezing avoidance by supercooling of tissue water in vegetative and reproductive structure of *Juniperus virginiana*. In *Plant Cold Hardiness and Freezing Stress: Mechanisms and Crop Implications*, vol.2. ed. P.H. Li & A. Sakai, pp. 367–77. New York: Academic Press.

George, M.F., Burke, M.J., Pellett, H.M. & Johnson, A.G. (1974). Low temperature exotherms and woody plant distribution. *HortScience*, **9**, 519–22.

Gholz, H.L. Fitz, F.K. & Waring, R.H. (1976). Leaf area difference associated with old-growth forest communities in the western Oregon Cascades. *Canadian Journal of Forest Research*, **6**, 49–57.

Grace, J. (1977). *Plant Response to Wind.* London: Academic Press.

Grace, J., Fasehun, F.E. & Dixon, M.A. (1980). Boundary layer conductance of the leaves of some tropical timber trees. *Plant, Cell and Environment*, **3**, 443–50.

Grace, J., Okali, D.U.U. & Fasehun, F.E. (1982). Stomatal conductance of two tropical trees during the wet season in Nigeria. *Journal of Applied Ecology*, **19**, 659–70.

Green, P.B., Erickson, R.O. & Buggy, J. (1971). Metabolic and physical control of cell elongation rate – *in vivo* studies in *Nitella. Plant Physiology*, **47**, 423–30.

Gregory, C.C. & Petty, J.A. (1973). Valve actions of bordered pits in conifers. *Journal of Experimental Botany*, **24**, 763–7.

Grier, C.C. & Running, S.W. (1977). Leaf area of mature north-western coniferous forests: relation to site water balance. *Ecology*, **58**, 893–9.

Grier, C.C. & Waring, R.H. (1974). Conifer foliage mass related to sapwood area. *Forest Science*, **20**, 205–6.

Gusta, L.V., Burke, M.J. & Kapoor, A.C. (1975). Determination of unfrozen water in winter cereals at subfreezing temperatures. *Plant Physiology*, **56**, 707–9.

Gusta, L.V., Fowler, D.B. & Tyler, N.J. (1982). Factors influencing hardening and survival in winter wheat. In *Plant Cold Hardiness and Freezing Stress: Mechanisms and Crop Implications*, vol.2. ed. P.H. Li & A. Sakai, pp. 23–40. New York: Academic Press.

Hadley, J.L. & Smith, W.K. (1983). Influence of wind exposure on needle desiccation and mortality for timberline conifers in Wyoming, USA. *Arctic and Alpine Research*, **15**, 127–35

Harbinson, J. & Woodward, F.I. (1984). Field measurements of gas exchange of woody plant species in simulated sunflecks. *Annals of Botany*, **53**, 841–5.

Heber, U., Krause, G.H., Schmitt, J.M., Klosson, R.J. & Santarius, K.A. (1981). Freezing damage to thylakoid membranes *in vitro* and *in vivo*. In *Effects of Low Temperatures on Biological Membranes*, ed. G.J. Morris & A. Clarke, pp. 263–83. London: Academic Press.

Holdridge, L.R. (1947). Determination of World plant formations from simple climatic data. *Science*, **105**, 367–8.

Holdridge, L.R. (1967). *Life Zone Ecology*, (rev.ed.). San José, Costa Rica: Tropical Science Center.

Houssard, C., Escarre, J. & Romane, F. (1980). Development of species diversity in some Mediterranean plant communities. *Vegetation*, **43**, 59–72.

Hsiao, T.C. (1973). Plant responses to water stress. *Annual Review of Plant Physiology*, **24**, 519–70.

Hsiao, T.C., Acevedo, E., Fereres, E. & Henderson, D.W. (1976). Stress metabolism: water stress, growth and osmotic adjustment. *Philosophical Transactions of the Royal Society, London. Series B*, **273**, 479–500.

Huber, B. & Schmidt, E. (1936). Weitere thermoelectrische Untersuchungen über den Transpirationstrom der Baume. *Tharandter Forstliches Jahrbuch*, **87**, 369–412.

Humboldt, A. von (1807). *Ideen zu einer Geographie der Pflanzen nebst einem Naturgemälde der Tropenländer*. Tübingen.

Humboldt, A. von & Bonpland, A. (1805). *Essai sur la Géographie des Plantes; Accompagne d'un Tableau Physique des Régions Equinoxiales*. Paris.

Jarvis, P.G. (1976). The interpretation of the variations in leaf water potential and stomatal conductance found in canopies in the field. *Philosophical Transactions of the Royal Society, London, Series B*, **273**, 593–610.

Jarvis, P.G. & Leverenz, J.W. (1983). Productivity of temperate, deciduous and evergreen forests. In *Encyclopedia of Plant Physiology*, vol. 12D, ed. O.L. Lange, P.S. Nobel, C.B. Osmond & H. Ziegler, pp. 233–80. Berlin: Springer-Verlag.

Jones, H.G. (1983). *Plant and Microclimate*. Cambridge University Press.

Jordan, W.R., Morgan, P.W. & Davenport, T.L. (1972). Water stress enhances ethylene-mediated leaf abscission in cotton. *Plant Physiology*, **50**, 756–8.

Kaku, S. (1971). Changes in supercooling and freezing processes accompanying leaf maturation in *Buxus*. *Plant and Cell Physiology*, **12**, 147–55.

Kira, T. (1975). Primary production of forests. In *Photosynthesis and Productivity in Different Environments*, ed. J.P. Cooper, pp. 5–40. Cambridge University Press.

Kira, T. (1978). Community architecture and organic matter dynamics in Tropical lowland rain forests of southeast Asia with special reference to Pasoh Forest, West Malaysia. In *Tropical Trees as Living Systems*, ed. P.B. Tomlinson & M.H. Zimmermann, pp. 560–90. Cambridge University Press.

Kira, T. & Shidei, T. (1967). Primary production and turnover of organic matter in different forest ecosystems of the western Pacific. *Japanese Journal of Ecology*, **17**, 70–87.

Kira, T., Shinozaki, K., & Hozumi, K. (1969). Structure of forest canopy as related to their primary productivity. *Plant and Cell Physiology*, **10**, 129–42.

Körner, C.H., Scheel, J.A. & Bauer, H. (1979). Maximum leaf diffusive conductance in vascular plants. *Photosynthetica*, **13**, 45–82.

Kozlowski, T.T. (1964). Shoot growth in woody plants. *Botanical Review*, **30**, 335–92.

Kozlowski, T.T (1976). Water supply and leaf shedding. In *Water Deficits and Plant Growth*, vol. IV, ed. T.T. Kozlowski, pp. 191–231. New York: Academic Press.

Kusumoto, T. (1957). Physiological and ecological studies on plant production in plant communities. 3. Ecological considerations of the temperature-photosynthesis curves of evergreen broad-leaved trees. *Japanese Journal of Ecology*, **7**, 126–30.

Lamb, H.H. (1982). *Climate, History and the Modern World.* London: Methuen.

Landsberg, J.J. & James, G.B. (1971). Wind profiles in plant canopies: studies on an analytical model. *Journal of Applied Ecology*, **8**, 729–42.

Larcher, W. (1980). *Physiological Plant Ecology*, 2nd edn. Berlin: Springer-Verlag.

Larcher, W. (1981*a*). Low temperature effects on Mediterranean sclerophylls: an unconventional viewpoint. In *Components of Productivity of Mediterranean Climate Regions – Basic and Applied Aspects*, ed. N.S. Margaris & H. A. Mooney, pp. 259–66. The Hague: Junk.

Larcher, W. (1981*b*). Resistenzphysiologische Grundlagen der evoltiven Kälteakklimatisation von Sprosspflanzen. *Plant Systematics and Evolution*, **137**, 145–80.

Larcher, W. (1982). Typology of freezing phenomena among vascular plants and evolutionary trends in frost acclimation. In *Plant Cold Hardiness and Freezing Stress: Mechanisms and Crop Implications*, vol.2, ed. P.H. Li & A. Sakai, pp. 417–26. New York: Academic Press.

Larcher, W. & Bauer, H. (1981). Ecological significance of resistance to low temperature. In *Encyclopedia of Plant Physiology*, vol. 12A, ed. O.L. Lange, P.S. Nobel, C.B. Osmond, & H. Ziegler, pp. 403–37. Berlin: Springer-Verlag.

Larcher, W. & Winter, A. (1981). Frost susceptibility of palms: experimental data and their interpretation. *Principes*, **25**, 143–52.

Larcher, W., Heber, U. & Santarius, K.A. (1973). Limiting temperatures for life functions. In *Temperature and Life*, ed. H. Precht, J. Christophersen, H. Hensel & W. Larcher, pp.195–292. Berlin: Springer-Verlag.

Levitt, J. (1980). *Responses of Plants to Environmental Stresses*, vol.1. *Chilling, Freezing and High Temperature Stresses*, 2nd edn. New York: Academic Press.

Leyton, L., Reynolds, E.R.C. & Thompson, F.B. (1967). Rainfall interception in forest and moorland. In *International Symposium on Forest Hydrology*, pp. 163–78. New York: Pergamon Press.

Li, P.H. & Sakai, A. (1978). (eds). *Plant Cold Hardiness and Freezing Stress: Mechanisms and Crop Implications.* New York: Academic Press.

Li, P.H. & Sakai, A. (1982) (eds). *Plant Cold Hardiness and Freezing Stress: Mechanisms and Crop Implications*, vol.2. New York: Academic Press.

Lieth, H. (1974) (ed). *Phenology and Seasonality Modelling.* Berlin: Springer-Verlag.

Lockhart, J.A. (1965). An analysis of irreversible plant cell elongation. *Journal of Theoretical Biology*, **8**, 264–76.

Lockwood, J.G. (1979). *Causes of Climate.* London: E. Arnold.

Lyons, J.M. (1973). Chilling injury in plants. *Annual Review of Plant Physiology*, **24**, 445–66.

Lyons, J.M., Graham, D. & Raison, J.K. (1979*a*). *Low Temperature Stress in Crop Plants. The Role of the Membrane.* New York: Academic Press.

Lyons, J.M., Raison, J.K. & Steponkus, P.L. (1979*b*). The plant membrane in response to low temperature: an overview. In *Low Temperature Stress in Crop Plants. The Role of the Membrane*, ed. J.M. Lyons, D. Graham, & J. K. Raison, pp. 1–24. New York: Academic Press.

McMichael, B.L., Jordan, W.R. & Powell, R.D. (1973). Abscission processes in cotton: induction by plant water deficit. *Agronomy Journal*, **65**, 202–4.

McNaughton, K.G. & Jarvis, P.G. (1983). Predicting effects of vegetation changes on transpiration and evaporation. In *Water Deficits and Plant Growth*, vol. VII, ed. T.T. Kozlowski, pp. 1–47. New York: Academic Press.

Maksymowych, R. (1973). *Analysis of Leaf Development*. Cambridge University Press.

Massman, W.J. (1982). Foliage distribution in old-growth coniferous tree canopies. *Canadian Journal of Forest Research*, **12**, 10–17.

Medway, Lord. (1972). Phenology of a tropical rain forest in Malaya. *Biological Journal of the Linnean Society*, **4**, 117–46.

Menaut, J.C. & Cesar, J. (1982). The structure and dynamics of a West African savannah. In *Ecology of Tropical Savannahs*, ed. B.J. Huntley & B.H. Walker, pp. 80-100. Berlin: Springer-Verlag.

Merrill, E.D. (1945). *Plant Life of the Pacific World*. New York: Macmillan.

Meryman, H.T., Williams, R.J. & Douglas, M.S.J. (1977). Freezing injury from solution effects and its prevention by natural or artificial cryoprotectants. *Cryobiology*, **14**, 287–302.

Meyen, F.J.F. (1846). *Outlines of the Geography of Plants: with Particular Enquiries Concerning the Native Country, the Culture, and the Uses of the Principal Cultivated Plants on which the Prosperity of Nations is Based*. London: Ray Society. (English translation of German edition of 1836.)

Miller, P.C. (1966). Leaf temperatures, leaf orientation and energy exchange in Quaking Aspen (*Populus tremuloides*) and Gambell's Oak (*Quercus gambellii*) in Central Colorado. *Oecologia Plantarum*, **2**, 241–70.

Miller, D.H. (1977). *Water at the Surface of the Earth. An Introduction to Ecosystem Hydrodynamics*. New York: Academic Press.

Monsi, M. (1968). Mathematical models of plant communities. In *Functioning of Terrestrial Ecosystems at the Primary Production Level*, pp. 131–49. Paris: Unesco.

Monsi, M. & Saeki, T. (1953). Über den Lichtfaktor in den Pflanzengesellschaften und seine Bedeutung für die Stoffproduktion. *Japanese Journal of Ecology*, **14**, 22–52.

Monteith, J.L. (1965). Evaporation and environment. In *The State and Movement of Water on Living Organisms*, ed. C.E. Fogg, pp. 205–34. Cambridge University Press.

Monteith, J.L. (1972). Solar radiation and productivity in tropical ecosystems. *Journal of Applied Ecology*, **9**, 747–66.

Monteith, J.L. (1973). *Principles of Environmental Physics*. London: E. Arnold.

Monteith, J.L. (1976) (ed). *Vegetation and the Atmosphere*, vol. II, *Case Studies*. London: Academic Press.

Monteith, J.L. (1977). Climate and the efficiency of crop production in Britain. *Philosophical Transactions of the Royal Society, London, Series B*, **281**, 277–94.

Monteith, J.L. & Elston, J. (1983). Performance and productivity of foliage in the field. In *The Growth and Functioning of Leaves*, ed. J.E. Dale & F.L. Milthorpe, pp. 499–518, Cambridge University Press.

Müller, M.J. (1982). *Selected Climatic Data for a Global Set of Standard Stations for Vegetation Science*. The Hague: Junk.

Nakagawara, S. & Sagisaka, S. (1984). Increase in enzyme activities related to ascorbic metabolism during cold acclimation in poplar twigs. *Plant and Cell Physiology*, **25**, 899–906.

Nienstaedt, H. (1974). Genetic variation in some phenological characteristics of forest trees. In *Phenology and Seasonality Modelling*. ed. H. Lieth, pp. 389–400. Berlin: Springer-Verlag.

Nomoto, N. (1964). Primary productivity of beech forest in Japan. *Japanese Journal of Botany*, **18**, 385–421.

Odum, S. (1979). Actual and potential tree-line in the North Atlantic region, especially in Greenland and the Faroes. *Holarctic Ecology*, **2**, 222–7.

Olsen, J.S. (1958). Rates of succession and soil changes on southern Lake Michigan sand dunes. *Botanical Gazette*, **119**, 125–70.

Ona, T.A. & Murata, N. (1981). Chilling susceptibility of the blue-green alga *Anacystis nidulans*. II. Stimulation of the passive permeability of cytoplasmic membrane at chilling temperatures. *Plant Physiology*, **67**, 182–7.

Oohata, S., & Sakai, A. (1982). Freezing resistance and thermal indices with reference to distribution of the Genus *Pinus*. In *Plant Cold Hardiness and Freezing Stress: Mechanisms and Crop Implications*. vol.2. ed. P.H. Li & A. Sakai, pp. 437–46. New York: Academic Press.

Paleg, L.G. & Aspinall, D. (1981) (eds). *The Physiology and Biochemistry of Drought Resistance in Plants*. Sydney: Academic Press.

Parker, J. (1963). Cold resistance in woody plants. *Botanical Review*, **29**, 123–201.

Paton, D.M. (1982). A mechanism for frost resistance in *Eucalyptus*. In *Plant Cold Hardiness and Freezing Stress: Mechanisms and Crop Implications*, vol.2,. ed. P.H. Li & A. Sakai, pp. 77–92. New York: Accademic Press.

Patterson, B.D., Graham, D. & Paull, R. (1979). Adaptation to chilling: survival, germination, respiration and protoplasmic dynamics. In *Low Temperature Stress in Crop Plants: The Role of the Membrane*, ed. J.M. Lyons, D. Graham & J.K. Raison, pp. 25–35. New York: Academic Press.

Patterson, B.D., Paull, R. & Smillie, R.R. (1978). Chilling resistance in *Lycopersicon hirsutum* Humb & Bonpl., a wild tomato with a wide altitudinal distribution. *Australian Journal of Plant Physiology*, **5**, 609–17.

Penman, H.L. (1948). Natural evaporation from open water, bare soil and grass. *Proceedings of the Royal Society of London, Series A*, **193**, 120–45.

Penman, H.L. (1963). Vegetation and hydrology. *Commonwealth Bureau of Soils Technical Communication*, **53**. Harpenden.

Pisek, A. & Winkler, E. (1958). Assimilationsvermögen und Respiration der Fichte (*Picea excelsa* Link) in verschiedener Höhenlage und der Zierbe (*Pinus cembra* L.) an der alpinen Waldgrenze. Planta, **51**, 518–43.

Polunin, N. (1960). *Introduction to Plant Geography and Some Related Sciences*. London: Longmans.

Potter, J.R. & Jones, J.W. (1977). Leaf area partitioning as an important factor in growth. *Plant Physiology*, **59**, 10–14

Priestley, C.H.B. & Taylor, R.J. (1972). On the assessment of surface heat flux and evaporation using large-scale parameters. *Monthly Weather Review*, **100**, 81–92.

Quinn, P.J. & Williams, W.P. (1978). Plant lipids and their role in membrane function. *Progress in Biophysics and Molecular Biology*, **34**, 109–73.

Raison, J.K., Chapman, E.A., Wright, I.C. & Jacobs, S.W.L. (1979). Membrane lipid transitions: their correlation with the climatic distribution of plants. In *Low Temperature Stress in Crop Plants: The Role of the Membrane*, ed. J.M. Lyons, D. Graham & J.K. Raison, pp. 177-86. New York: Academic Press.

Rajashekar, C. & Burke, M.J. (1978). The occurrence of deep undercooling in the Genera, *Pyrus*, *Prunus* and *Rosa*: a preliminary report. In *Plant Cold Hardiness and Freezing Stress: Mechanisms and Crop Implications*, ed. P.H. Li & A. Sakai, pp.213–25. New York: Academic Press.

Rajashekar, C., Gusta, L.V. & Burke, M.J. (1979). Membrane structural transitions: probable relation to frost damage in hardy herbaceous species. In *Low Temperature Stress in Crop Plants. The Role of the Membrane*, ed. J.M. Lyons, D. Graham & J.K. Raison, pp.255-74. New York: Academic Press.

Rasmussen, D.H. & MacKenzie, A.P. (1972). Effect of solute on ice-solution interfacial free energy: calculation from measured homogeneous nucleation temperatures. In *Water Structure at the Water Polymer Interface*, ed. H.H.G. Jellinok, pp.126–45. New York: Plenum Publishing Corporation.

Rasmusson, E.M. (1971). A study of the hydrology of eastern North America using atmospheric flux data. *Monthly Weather Review*, **99**, 119–35.

Rauner, Iu.L. (1972). *Teplovoi Balans Rastitel'nogo Pokrova*. Leningrad: Gidrometeoizdot.

Rosenberg, N.J. (1974). *Microclimate: the Biological Environment*. New York: Wiley.

Running, S.W., Waring, R.H. & Rydell, R.A. (1975). Physiological control of water flux in conifers: a computer simulation model. *Oecologia*, **18**, 1-16.

Rutter, A.J., Morton, A.J. & Robins, P.C. (1975). A predictive model of rainfall interception in forests. II. Generalisation of the model and comparison with observations in some coniferous and hardwood stands. *Journal of Applied Ecology*, **12**, 367–80.

Sakai, A. (1978). Freezing tolerance of evergreen and deciduous broad-leaved trees in Japan with reference to tree regions. *Low Temperature Science, Series B*, **36**, 1–19.

Sakai, A. (1979). Freezing avoidance mechanism of primordial shoots of conifer buds. *Plant and Cell Physiology*, **20**, 1381–90.

Sakai, A. (1983). Comparative studies on freezing resistance of conifers with special reference to cold adaptation and its evolutive aspects. *Canadian Journal of Botany*, **61**, 2323–32.

Sakai, A. & Wardle, P. (1978). Freezing resistance of New Zealand trees and shrubs. *New Zealand Journal of Ecology*, **1**, 51–61.

Sakai, A. & Weiser, C.J. (1973). Freezing resistance of trees in North America with reference to tree regions. *Ecology*, **54**, 118–26.

Santarius, K.A. (1984). Effective cryoprotection of thylakoid membranes by ATP. *Planta*, **161**, 555–61

Satoo, T. (1970). A synthesis of studies by the harvest method: primary production relations in the temperate deciduous forests of Japan. In *Analysis of Temperate Forest Ecosystem*, ed. D.E. Reichle, pp. 55–72. London: Chapman & Hall.

Satoo, T. (1983). Temperate broad-leaved evergreen forests of Japan. In *Ecosystems of the World*, 10, *Temperate Broad-Leaved Evergreen Forests*, ed. J.D. Ovington, pp. 169–89.

Schimper, A.F.W. (1898). *Pflanzengeographie auf physiologischer Grundlage*. Jena.

Schnelle, F., Baumgartner, A. & Freitag, E. (1984). *Arboreta Phaenologica*. **28**.

Schouw, J.F. (1823). *Grunzüge einer allgemeinen Pflanzengeographie*. Berlin.

Schulze, E.-D. (1982). Plant life forms and their carbon, water and nutrient relations. In

Encyclopedia of Plant Physiology, vol. 12B, ed. O.L. Lange, P.S. Nobel, C.B. Osmond & H. Ziegler, pp. 616–76.

Schulze, E.D. & Hall, A.E. (1982). Stomatal responses, water loss and CO_2 assimilation rates of plants in contrasting environments. In *Encyclopedia of Plant Physiology*, vol. 12B, ed. O.L. Lange, P.S. Nobel, C.B. Osmon & H. Ziegler, pp. 181–230.

Senser, M. & Beck, E. (1977). On the mechanisms of frost injury and frost hardening of spruce chloroplasts. *Planta*, **137**, 195–201.

Senser, M., Dittrich, P., Kandler, O., Thanbichler, A. & Kuhn, B. (1971). Isotopenstudien über den Einfluss der Jahreszeit auf den Oligosaccharidumsatz bei Coniferen. *Berichte der Deutschen Botanischen Gesellschaft*, **84**, 445–55.

Senser, M., Schötz, F. & Beck, E. (1975). Seasonal changes in structure and function of spruce chloroplasts. *Planta*, **126**, 1–10.

Siau, J.F. (1971). *Flow in Wood.* Syracuse, New York.

Spurr, S.H. & Barnes, B.V. (1980). *Forest Ecology*, 3rd edn. New York: Wiley.

Taylor, F.G. Jnr (1974). Phenodynamics of production in a mesic deciduous forest. In *Phenology and Seasonality Modelling*, ed. H. Lieth, pp. 237–54. Berlin: Springer-Verlag.

Tieszen, L.I. (1978) (ed). *Vegetation and Production Ecology of an Alaskan Arctic Tundra.* New York: Springer-Verlag.

Tieszen, L.I., Lewis, M.C., Miller, P.C., Mayo, J., Chapin, III, F.S. & Oechel, W. (1981). An analysis of processes of primary production in tundra growth forms. In *Tundra Ecosystems: a Comparative Analysis*, ed. L.C. Bliss, O.W. Heal & J.J. Moore, pp. 285–356. Cambridge University Press.

Tranquillini, W. (1979). *Physiological Ecology of the Alpine Timberline*. Berlin: Springer-Verlag.

Tyree, M.T., Jarvis, P.G. (1982). Water in tissues and cells. In *Encyclopedia of Plant Physiology*, vol. 12B, ed. O.L. Lange, P.S. Nobel, C.B. Osmond & H. Ziegler, pp. 35–77.

Vareschi, V. (1980). *Vegetationsökologie der Tropen.* Stuttgart: Eugen Umer.

Volger, H.G. & Heber, U. (1975). Cryoprotective leaf proteins. *Biochimica et biophysica acta*, **412**, 335–49.

Walker, B.H. & Noy-Meir, I. (1982). Aspects of the stability and resilience of savannah ecosystems. In *Ecology of Tropical Savannahs*, ed. B.J. Huntley & B.H. Walker, pp. 556–90. Berlin: Springer-Verlag.

Walter, H. (1931). *Die Hydratur der Pflanzen und ihre physiologisch-ökologische Bedeutung*. Jena: Gustav-Fischer-Verlag.

Walter, H. (1939). Grasland, Savanne und Busch der ariden Teile Afrikas in ihrer ökologischen Bedingtheit. *Jahrbücher für Wissenschaftliche Botanik*, **87**, 750–860.

Walter, H. (1968). *Die Vegetation der Erde in öko-physiologischer Betrachtung*. vol. II. *Die Gemässigten und arktischen Zonen*. Jena: Gustav-Fischer-Verlag.

Walter, H. (1973). *Die Vegetation der Erde in öko-physiologischer Betrachtung*. vol.I. *Die tropischen und subtropischen Zonen*, 3rd edn. Stuttgart: Gustav-Fischer-Verlag.

Walter, H. (1976). *Die ökologischen Systeme der Kontinente* (*Biogeosphäre*)*: Prinzipien ihrer Gliederung mit Beispielen*. Stuttgart: Gustav-Fischer-Verlag.

Walter, H. (1979). *Vegetation of the Earth and Ecological Systems of the Geo-Biosphere*, 2nd edn. New York: Springer-Verlag.

Wardle, P. (1971). An explanation for alpine timberline. *New Zealand Journal of Botany*, **9**, 371–402.

Wardle, P. (1977). Plant communities of Westland National Park (New Zealand) and neighbouring lowland and coastal areas. *New Zealand Journal of Botany*, **15**, 323–98.

Wareing, P.F. (1956). Photoperiodism in woody plants. *Annual Review of Plant Physiology*, **7**, 191–214.

Waring, R.W. & Franklin, J.F. (1979). Evergreen coniferous forests of the pacific Northwest. *Science*, **204**, 1380–6

Waring, R.H. & Running, S.W. (1978). Sapwood water storage: its contribution to transpiration and effect upon water conductance through the stems of old-growth Douglas-Fir. *Plant, Cell and Environment*, **1**, 131–40.

Waring, R.H., Emmingham, W.H., Gholz, H.L. & Grier, C.C. (1978). Variation in maximum leaf area of coniferous forests in Oregon and its ecological significance. *Forest Science*, **24**, 131–40.

Waring, R.H., Rogers, J.J. & Swank, W.T. (1981). Water relations and hydraulic cycles. In *Dynamic Properties of Forest Ecosystems*, ed. D.E. Reichle, pp. 205–64. Cambridge University Press.

Waring, R.H., Whitehead, D. & Jarvis, P.G. (1979). The contribution of stored water to transpiration in Scots pine. *Plant, Cell and Environment*, **2**, 309–17.

Webber, P.J. & Klein, D.R. (1977). Geobotanical and ecological observations at two locations in the west–central Siberian arctic. *Arctic and Alpine Research*, **9**, 305–15.

Whitehead, D. & Jarvis, P.G. (1981). Coniferous forests and plantations. In *Water Deficits and Plant Growth*, vol. VI, ed. T.T. Kozlowski, pp. 49–152. New York: Academic Press.

Woodward, F.I. (1975). The climatic control of the altitudinal distribution of *Sedum rosea* (L.) Scop. and *S. telephium* L. II. The analysis of plant growth in controlled environments. *New Phytologist*, **74**, 335–48.

Woodward, F.I. & Pigott, C.D. (1975). The climatic control of the altitudinal distribution of *Sedvm rosea* (L.) Scop. and *S. telephium* L. I. Field observations. *New Phytologist*, **74**, 323–34.

Woodward, F.I. & Sheehy, J.E. (1983). *Principles and Measurements in Environmental Biology*. London: Butterworths.

Wulff, E.V. (1943). *An Introduction to Historical Plant Geography* (English translation of Russian edition, Chronica Botanica, Waltham, Mass.).

Ylimaki, A. (1962). The effect of snow cover on temperature conditions in the soil and overwintering of field crops. *Annales Agriculturae Fenniae*, **1**, 192–216.

Yoshie, F. & Sakai, A. (1982). Freezing resistance of temperate deciduous forest plants in relation to their life form and microhabitat. In *Plant Cold Hardiness and Freezing Stress: Mechanisms and Crop Implications*, vol. 2, ed. P.H. Li & A. Sakai, pp. 427–36. New York: Academic Press.

Ziegler, P. & Kandler, O. (1980). Tonoplast stability as a critical factor in frost injury and hardening of spruce (*Picea abies* L. Karst.) needles. *Zeitschrift für Pflanzenphysiologie*, **105**, 229–39.

Zimmerman, M.H. (1964). Effect of low temperature on ascent of sap in trees. *Plant Physiology*, **39**, 568–72.

Zinke, P.J. (1967). Forest interception studies in the United States. In *Forest Hydrology*, ed. W.E. Sopper & H.W. Lull, pp. 137–60. New York: Pergamon Press.

Zubenok, L.I. (1974). Evaporation deficit under various climatic conditions on land. *Soviet Hydrology*, **13**, 251–7.

5

Climate and the distribution of taxa

'No room! No room!' they cried out when they saw Alice coming. 'There's plenty of room!' said Alice.
L. Carroll.

Introduction

The previous chapter outlined an attempt to define the major global zonations of vegetation by an objective technique based on ecophysiological responses to temperature and on the balance between evaporation and precipitation. Even these large-scale considerations show substantial ecophysiological differences between species common to a particular vegetation type. It follows therefore that these differences may be critical in controlling distribution within a vegetational zone. This chapter aims to investigate the mechanisms by which climate may control such variations in the distribution of plant taxa and, in particular, considers the dynamic aspects of plant responses in terms of dispersal and in the ability to occupy space.

Global perspective

The emphasis in the previous chapter was one of a climatic limitation to the polar spread of species, determined by their ability to survive critical, cardinal temperatures. The prediction from these discussions is that taxonomic diversity should decline in a poleward direction. This hypothesis has been tested by investigating latitudinal variation in a number of families of angiosperms (from Heywood, 1979) and arboreal gymnosperms (from Hora, 1981). A total of 313 families were included in the analysis. The distribution of the families in latitudinal bands of 15 ° is shown on Fig. 5.1. For both the northern and southern hemispheres family diversity is greatest near the equator, declining markedly from latitudes 30 ° to 90 °. The same pattern may also be seen for species diversity (Rejmánek, 1976).

The absolute minimum temperature for each of the latitudinal bands has been extracted from the climatic data presented in Müller (1982) and is plotted against family number on Fig. 5.2. There is a strong correlation between family diversity and absolute minimum temperature. The

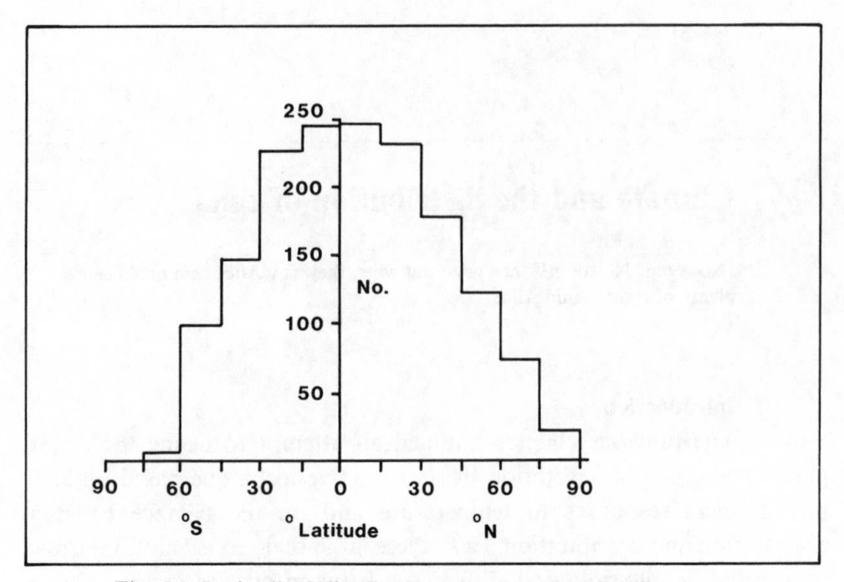

Fig. 5.1. Latitudinal distribution of major plant families.

regression line shown in the figure suggests a reduction of 3.3 families per °C reduction in minimum temperature. Such a correlation is predictable from the evidence of the previous chapter. However, it remains at best a correlation, especially when it is considered that these mean values cover a wide range of stations differing in altitude, continentality and precipitation.

An intriguing feature to emerge from the correlation is the way in which the regression line differentiates between the northern and southern hemispheres when the minimum temperature is below 0 °C. It is interesting to consider whether this is an 'artefact' derived in the averaging process which produces the mean values, or whether it has some significance to the control of plant diversity. The mean minimum temperature will be a function of the frequency and topographical and altitudinal diversity of the meteorological stations selected for the analysis, but a single mean value will clearly mask the range of diversity within a latitudinal range. In the same way, the number of families may also be a function of habitat diversity. If this is the case then larger areas of land should have a greater probability than smaller areas of presenting the complete spectrum of variation which might be required for the differing habitat requirements of different families (Preston, 1960; MacArthur, 1965; MacArthur & Wilson, 1967; Connor & McCoy, 1979). The areas of land included in the different latitudinal ranges are shown in Table 5.1.

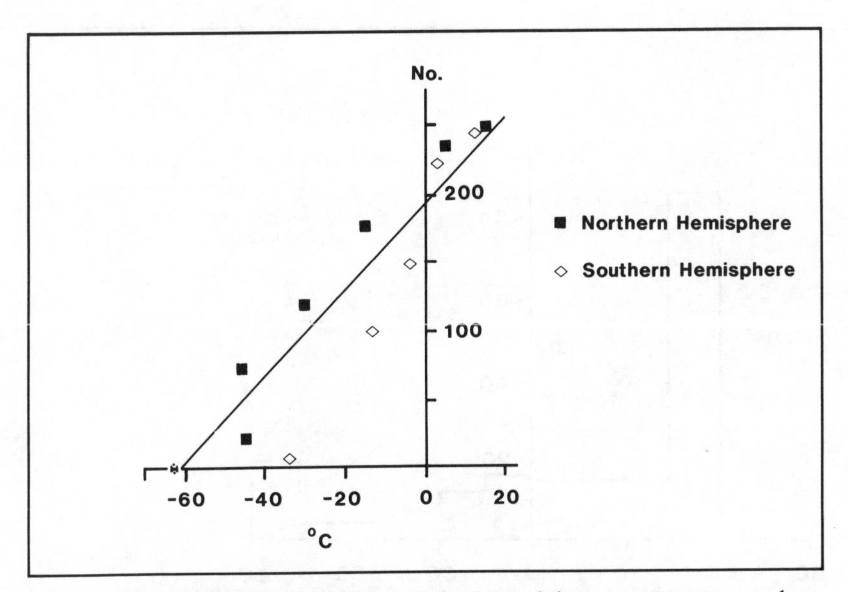

Fig. 5.2. Relationship between absolute minimum temperature and family number at different latitudes.

Table 5.1. *Approximate terrestrial surface areas of the globe*

Latitudinal range	Area
	[Mm² (10⁶ Km²)]
0–15 ° N	14.4
15–30 ° N	19.4
30–45 ° N	19.8
45–60 ° N	24.6
60–75 ° N	14.8
75–90 ° N	1.2
0–15 ° S	14.9
15–30 ° S	14.3
30–45 ° S	9.3
45–60 ° S	1.1
60–75 ° S	4.8

The terrestrial surface area of the northern hemisphere is considerably more extensive than the southern hemisphere at latitudes exceeding 15 °. It is possible therefore that the variation shown on Fig. 5.2 is simply a measure of hemispherical differences in surface area. Some measure of habitat diversity may be realised by calculating the number of families per unit of land surface, the family area. Considerations of the poleward decline of diversity shown on Fig. 5.1 suggest that, irrespective of habitat

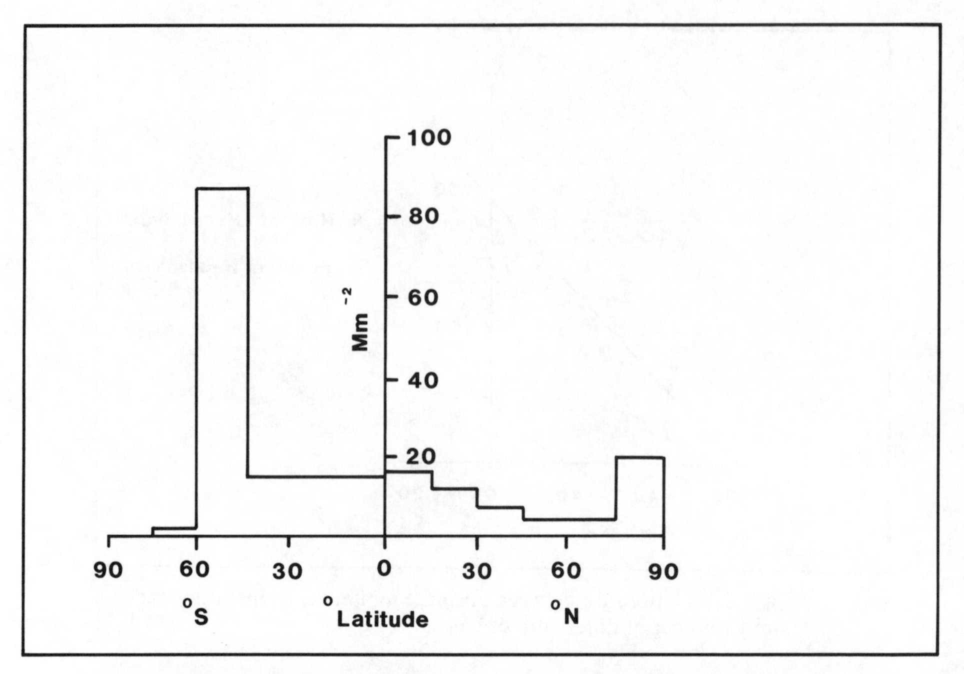

Fig. 5.3. Latitudinal variations in the number of families per unit area.

diversity, family area might decrease with latitude, reflecting the reduced diversity resulting from the selective forces of low temperatures. The latitudinal range of family area has been presented in Fig. 5.3.

The latitudinal change in family area is not simply monotonic, although in the northern hemisphere a clear reduction in family area progresses to 75° N. In both hemispheres the family area increases markedly at the margins of the major land masses between 75–90° N and 45–60° S. The area of land between 60° and 75° S with a low family area is well isolated from any major continent, perhaps reflecting the limitations of plant dispersal across ocean surfaces. Certainly this latitudinal zone is much less diverse than the equivalent zone in the northern hemisphere.

The explanation for the two peaks in family area is not clear. Both latitudinal ranges have virtually identical surface areas (1.1 to 1.2 Mm^2) and, particularly in the southern hemisphere, are the dwindling expanses of large continents at lower latitudes.

Within the latitudinal range of 45° to 60° S, which lies in the main at the foot of South America, 93% of the families are also common to the lower latitudes of South America. Of the remaining 7%, 6% are found in Australasia and 1% in South Africa, possibly implying some historical commonality relating to the ancient Gondwanaland connection between these different areas. In fact, 21% of the families in South America between 45° and 60° S have their geographically closest neighbours in either South Africa or Australasia, or both, providing some quantitative measure of the relevance of at least 60 million years of history to the present day distribution of families. This historical connection may be an important feature of the diversity in the latitudinal range of 45–60° S.

Diversity in the northern latitudes is lower than in the south and 91% of the families are truly cosmopolitan compared with 68% in the southern range. Land in the latitudes from 75° to 90° N is confined to three areas: part of Greenland, Ellesmere Island and the Queen Elizabeth Island to the north and north east of Canada; Svalbard in the Arctic Ocean and the north of Novaya Zemlya and the Taymyr Peninsula of the USSR. The majority of the land area consists of Greenland and the islands to the north of Canada. However only 57% of the families occur in this area. Svalbard has the smallest area, containing only 43% of the species. The Taymyr Peninsula, which is comparable in area to Svalbard is, in contrast, floristically richer, containing 96% of the families present throughout this range of latitudes. In these cases, unlike the southern latitudes, it is not possible to be as unequivocal in ascribing historical bases to present distribution because the families are particularly widespread. However the limited diversity of the Islands compared with the Taymyr Peninsula at the tip of the vast Eurasian continent suggests the crucial importance of the probability of plant dispersal to diversity. Viewed in these terms it is possible to suggest that the diversity in the latitudes between 45° and 60° S and 75° and 90° N is due largely to dispersal-efficient contact with the large diversity of large continents. Once this efficiency of dispersal is reduced then diversity is also reduced, as in Svalbard or in the latitudes between 60° and 75° S.

The high diversity in these selected areas, measured as family area, implies that much of the ground area in the adjacent continents is redundant in terms of diversity.

Dispersal and migration

It appears, therefore, that investigations on the manner in which climate controls the distribution of taxa must be concerned with the rate and extent to which the dispersal of different species responds to changes in climate. Indeed, it might be considered that the response of species to the sorts of changes in climate described in Chapter 2 can only be understood as the combined phenomena of the dispersal or spatial spread of plant propagules and the migration of plant populations. It is unlikely that the two processes can be studied independently because of the

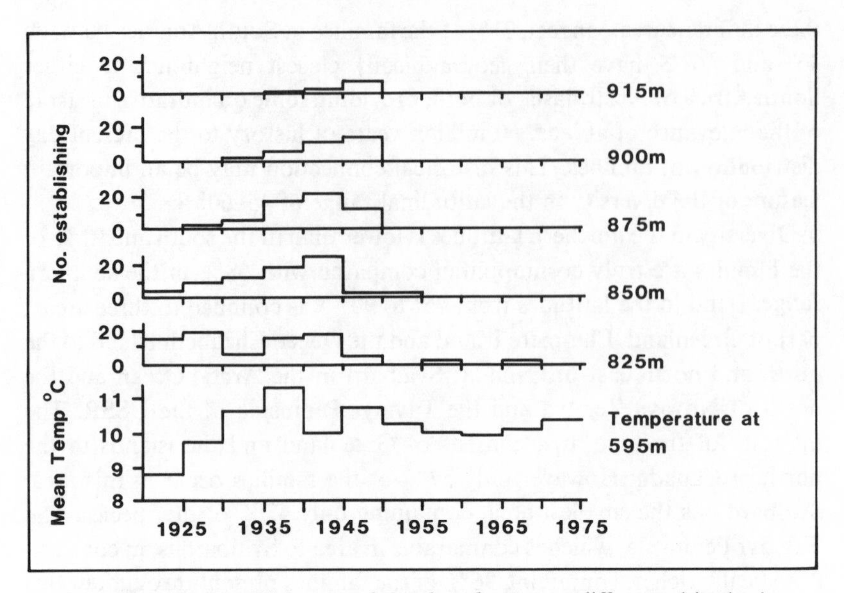

Fig. 5.4. Establishment of *Betula pubescens* at different altitudes in Sweden.

random nature of the actual dispersal process. In addition, the very nature of the spread of a species implies an appreciable period of time for the process to occur (Chapter 2), therefore historical ecology must be the approach to describe this response to climate. These responses will be described below, but will be followed by more mechanistic or ecophysiological explanations of why the ranges of some species may expand whilst those of others may not, or may even contract.

One of the most productive areas for the investigation of migration is at the limits to a particular area of distribution. Such a limit may be very sensitive to changes in climate if a threshold phenomenon is the cause of the limit and can provide evidence for the mechanisms by which species adjust their range. Kullman (1979, 1983) has provided careful analyses of the variations in the tree and forest limits of birch, *Betula pubescens*, in central Sweden since the turn of the century. This work is also particularly interesting in that it re-enumerated the same area that Smith (1920) had surveyed in the period between 1915 and 1916. The evidence which is presented therefore provides an accurate assessment of any changes in the tree limit of birch over a period of about 60 years.

Kullman (1979) demonstrated that the altitude of the 1915–16 tree limit had increased in about 75% of the localities that he investigated, with the remaining 25% being unchanged. He concluded that the rise in the tree limit was caused by the increased frequency of warm summers during the

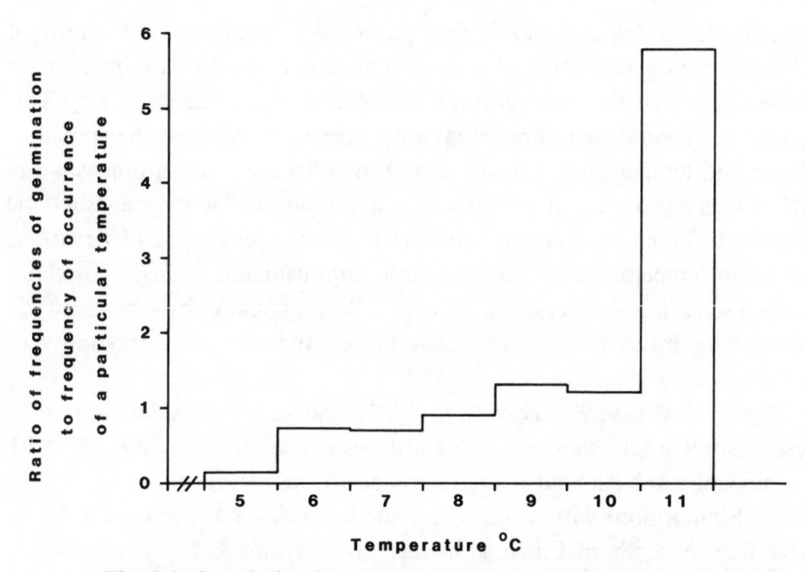

Fig. 5.5. Correlation between temperature and seed germination and establishment of *Betula pubescens*.

period of 1930 to 1949. A comparison between the mean annual temperature near this region (Storlien at 595 m) and dated establishment of birch, at and above the 1915–16 tree limit at 825 m to the present tree limit at 915 m, are shown in Fig. 5.4.

Birch establishment occurred well above the original tree limit, although decreasing significantly with altitude. A climatic explanation of this expansion in the range of birch is encouraged by the strong correlation between the high mean temperatures and establishment, and the sharp cut-off in establishment in the period of 1950–55 when temperature fell and remained cool to 1975.

If a typical adiabatic lapse rate (for Europe) of $-0{\cdot}67$ °C 100 m^{-1} of altitude (Woodward & Pigott, 1975) is applied to the climatic data for Storlien (595 m), then it is possible to estimate the mean temperature at the different altitudes. A linear regression and correlation analysis between the 5–year mean values for temperature, calculated in this way, and establishment, is less than revealing with a particularly insignificant correlation coefficient of 0.055. These data do not therefore provide an explanation of establishment in terms of temperature. This may of course be because the data are for 5 year intervals, hiding more significant annual correlations. Fortunately, the data of Kullman (1979) provide dates of establishment for the majority of the plants that were investigated. A sub-sample of 130 of these measurements has been taken, the dates of establishment determined and then plotted as a histogram of the ratio of the numbers germinating at a particular temperature to the number of observations of that temperature, corrected for altitude (Fig. 5.5). This technique should therefore reveal any correlation between the spread of birch and temperature. The mean altitude of these observations was 824 m, with a maximum at 940 m and a minimum at 730 m; the adiabatic lapse rate for temperature with altitude has been applied to the observations of mean temperature at Storlien. Birch establishment strongly correlates with increasing temperatures above 5 °C and markedly above 10 °C. Increasing mean temperature does indeed increase the probability of establishment.

Agren, Isaksson & Zackrisson (1983) also show a similar pattern of establishment for *Pinus sylvestris* and *Picea abies* in northern Sweden, with significant establishment being restricted to the period of about 1910 to 1970. Similar observations have been made for *Larix laricina* at the Arctic tree line in northern Canada. In this case Payette & Lajeunesse (1980) noted the establishment and extension in the range of larch during the warmer periods of the twentieth century. However in the later, cooler periods of the 1960s and 1970s, adult individuals suffered severe mortality, suggesting a subsequent contraction of range dependent on the response of mature individuals. Kullman (1983), on the other hand, points out the reverse situation for *Betula pubescens*, with adult individuals surviving, with minimal mortality, in the cooler years when there is no establishment from seed. As Kullman (1979, 1983) has pointed out, it is clear that an understanding of the ecophysiological responses of different stages of the life cycle are crucial in understanding the control of plant distribution.

Kullman (1983) also notes that whereas birch was able to increase its geographical range during the climatic amelioration in the period from 1930 to 1950, the same was not true for either *Pinus sylvestris* or *Picea abies*, although recruitment from seed was observed adjacent to mature individuals. Kullman maintains, on the basis of ^{14}C-dating of pine, that the species is in a distributional recession, albeit long-term because of the longevity of the species (approximately 400 years). He considers that the recesssion is caused by increasing snow fall in the area, which is more adverse for the establishment of pine than for birch. As with all correlations, however, other factors such as those related to human activity cannot be discounted.

It is interesting to note the strong effect on population dynamics of variations in the plant response times (Chapter 2) to climate, both between the species, of birch and pine, and between different stages of the life cycle of one species (in particular, the seedling and adult stages of birch or pine).

The shorter response time of the spread of birch allows a response of its geographical range to short periods of climatic amelioration. The spread of mature individuals of pine, on the other hand, evidently has a longer response time and is unable to effect a change in range, even though seed germination with a short response time (Chapter 2) can respond.

The changes in annual mean temperature leading to, or at least correlating with, the apparently widespread phenomenon of an increase in establishment at the tree line are quite small (in the order of about 2 °C or less). In contrast, the change in temperature following the last ice age (see Chapters 1 and 2) was considerably larger. Mörner (1980), for example, has estimated the temperature of Lake Tingstade Trask, on the Swedish island of Gotland (in the Baltic Sea), using the oxygen isotopic technique described in Chapter 1. Dependent on the confidence in the estimates of temperature (see Chapter 1 and Mörner & Wallin, 1976) it appears that the mean temperature of the lake increased by about 15 °C in 1000 years after the last ice age. The effect of this climatic change on vegetation is inconceivable from present day ecology and so we must look to palaeoecology for the nature of the vegetational response (Chapter 1).

Evidence from palynological studies suggests that forest trees moved rapidly over the landscape for an appreciable period after the ice age, from about 12 000 BP to 6000 BP, with rates of migration in both North America and Europe estimated at an average of about 300 m yr^{-1} (Davis, 1978, 1981; Huntley & Birks, 1983). The response of vegetation to the considerable change in climate following the last glaciation was evidently swift in terms of migration, although lasting for a period of at least 6000 years. During this period the vegetation may not have been in equilibrium with the climate, which had changed rather little since about 9000 BP. It is not obvious, therefore, how to extract information easily on the response of plants to climate, when much of the obvious response is a delayed one to the earlier climatic amelioration.

A different approach to investigating the nature of plant dispersal and the impact of climate and habitat diversity on plant distribution can however be attempted by a combination of palaeoecological and present-day, phytogeographic information. Such an analysis will be attempted using the extensive information presented in the studies by Hulten (1960) on the distribution of vascular plants on the Aleutian and Commander Islands. These islands form a chain between the Kamtchatka peninsula of the USSR and the Alaskan peninsula of the United States of America (Fig. 5.6. from Tatewaki & Kobayashi, 1934). The aim here is to investigate the spread of species in both spatial and temporal terms: spatially, in terms of the sizes and distances between the islands; temporally, in terms of the response times of geographical spread following changes in climate which may be inferred from palaeoecological evidence.

Hulten has worked and published on the floras of the Kamtchatka Peninsula, to the east of the Aleutian Islands (1927–1928), the flora of the Aleutian Islands, with notes on the flora of the Commander Islands (1960) and the Flora of Alaska and Yukon (1941–1950). This broad base of floristic knowledge is quite ideal for describing the present-day distribution and geographical affinities of the Aleutian and Commander Islands. The flora of the Aleutian Islands also includes an appendix of maps showing the distribution of 533 species. About 70% of these species are found in both Kamtchatka and in Alaska. These species may be generally classified as arctic or tundra species and are species which are well able to survive the windy and cool climate of the islands (Hulten, 1960).

Of the remaining species, about 17% are found in Alaska but not Kamtchatka and about 11% are found in Kamtchatka but not Alaska. According to the floristic affinities of the islands (Tatewaki, 1963), the American species occur along the American Pacific coast and in North America itself, whilst those from the USSR are also found along the Pacific coast, although on the Asiatic side and in eastern Asia. These species are clearly from rather milder climates than the numerically dominant, arctic species.

The interest in these species in the present context is that their origins – i.e. either the Asian or American mainland – is fairly certain; therefore, present-day distributions will provide a measure of the ability of the species to 'island hop', or to migrate between islands. In addition to this measure of migration, the effects of island size and climatic heterogeneity may also be considered, providing a measure of the influence of local climate on the process of migration.

The number of these selected species on each of the major islands (Fig. 5.6) is shown on Fig. 5.7. Irrespective of the mainland origins of the species, number declines more or less geometrically from the assumed continental source. The line on each diagram assumes a probability of 81% that a species will migrate 100 km from its mainland source. This curve provides a general description of the observations, although it is clearly inaccurate in some cases. This general estimate of the probability of distant migration could only be achieved by the sort of analysis described here, and as a result of extensive field observations. Studies on the population dynamics of species, or on the actual process of dispersal, could not provide such information because of the over-

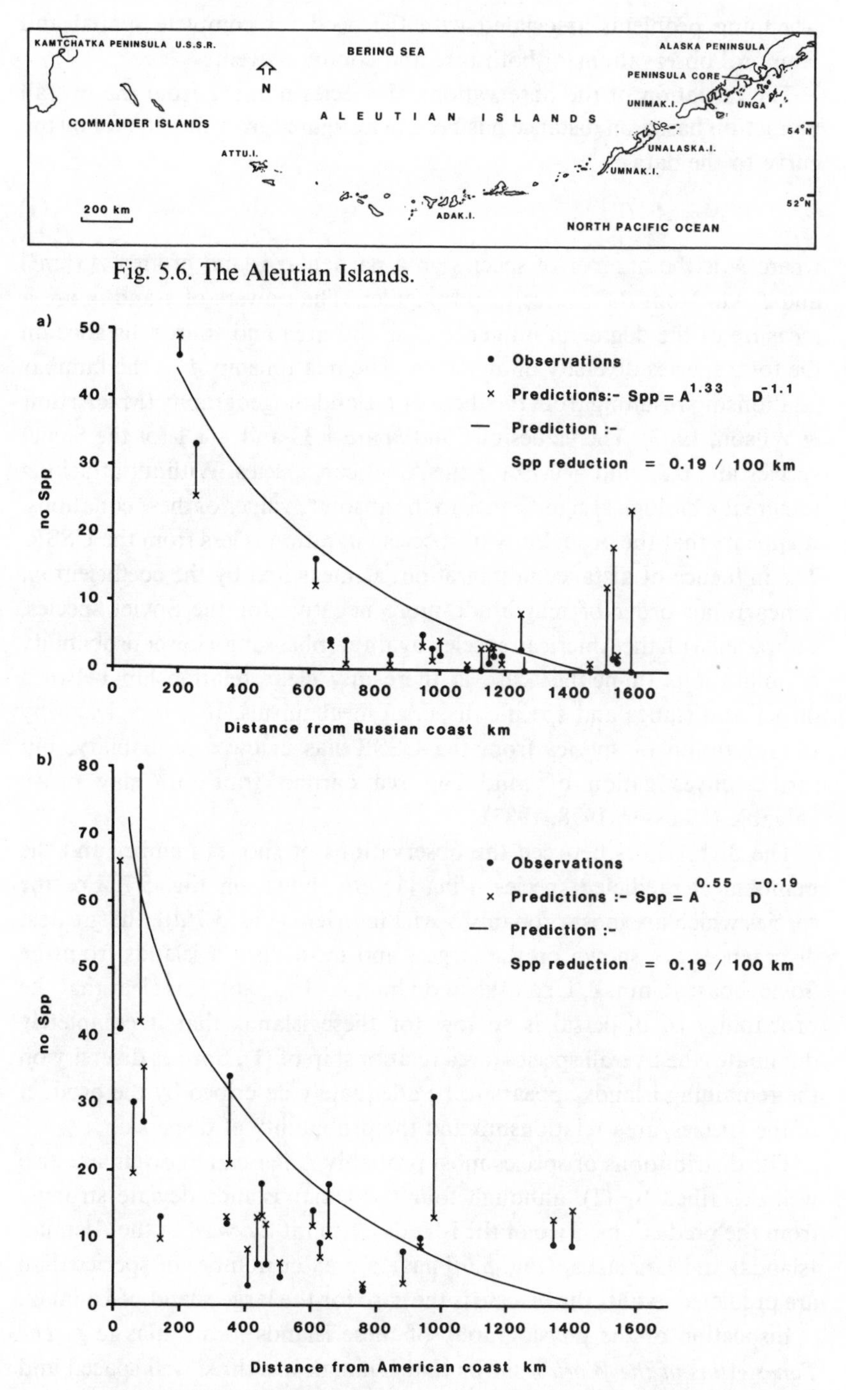

Fig. 5.6. The Aleutian Islands.

Fig. 5.7. Distribution of species on the Aleutian Islands. N, number of species on island; A, island area (km²); D, distance from seed source (km).

whelming problems associated with the need for complete spatial and temporal observations of both rare and common events.

The deviation of the observations of species number from the overall prediction based on distance has been investigated by fitting the following curve to the data:

$$N = A^z D^m \quad (1)$$

where N is the number of species on a particular island of area A (km²) and D km from the source of propagules. The powers of z and m are a measure of the degree of influence of island area and source distance on the total species diversity of an island. The relationship A^z is the familiar relationship resulting from the theory of island biogeography (MacArthur & Wilson, 1967). The values of z and m are 1.33 and -1.1 for the Soviet species and 0.55 and -0.19 for the American species. Without attaching too great a biological significance to the absolute values of these equations, it appears that the probability of species migration is less from the USSR. The influence of distance in migration, as measured by the coefficient m, is nearly an order of magnitude more negative for the Soviet species, compared with the American species, again emphasising a lower probability of migration. In neither case is there any clear relationship between dispersal distance and specific dispersal mechanisms. It is not clear why the migration of species from the USSR has a lower probability, but further investigation of wind and sea current transport may prove valuable (Heusser, 1978, 1983).

The divergences between the observations of species number and the numbers of predicted species using (**1**) are shown on Fig. 5.7. For the species which are most probably Soviet in origin (Fig. 5.7(*a*)), the greatest divergences are shown on the largest and most distant islands from the Soviet coast: Umnak, Unalaska and Unimak (Fig. 5.6). It is clear that the probability of dispersal is so low for these islands that it completely dominates the overall species/area relationship of (**1**). Species diversity on the remaining islands appears to be adequately described by the product of the species/area relationship and the probability of dispersal.

The distributions of species most probably American in origin are also well described by (**1**), although four particular islands deviate strongly from the predictions. Two of the islands [Attu (at the west of the Aleutian Islands) and Unalaska, (Fig. 5.6)] have a greater number of species than are predicted, whilst the reverse is the case for the large island of Unimak.

Inspection of the physiography of these islands in an atlas (e.g. *The Times Atlas of the World*) shows that Unimak has three well spaced and active volcanoes, with peaks of 1073 m, 2000 m and 2861 m. The highest volcano is also the centre of an extensive glacier. It is possible that extreme disturbance following volcanic activity may limit the migration of species, or the diversity of the island.

Unalaska has a more diverse flora and is slightly smaller than Unimak, with one volcano (2035 m) and a rather small glacier. Whereas the volcanoes on Unimak are spread around the island, that on Unalaska is at the northern end of the island, perhaps reducing the area affected after disturbance. Unalaska also possesses an active port, with potential for the introduction of species by humans. The island of Attu, also with a diverse flora, has no active volcano and no peak above 1000 m.

The implication from these observations is that the migration of species, or island diversity, may be a function of extremes of disturbance, in addition to any explanations relating to dispersal (and in the absence of any precise detail of the geological and edaphic variation between the islands). A number of the islands in the chain are also volcanic, so it is of interest to consider the effects of eruptions on plant distribution.

Such a study is possible as a result of the palaeoecological investigations by Heusser on a number of these islands (Fig. 5.6): the Alaska Peninsula and Unga (1983), Umnak (1973) and Adak (1978). Palaeoecological evidence from all sites indicates an active volcanic period from the end of the last glaciation on the Aleutian Islands, a time which varied from about 12 000 BP on Adak to about 7000 BP on Attu (C.J. Heusser, personal communication).

Three major volcanic events are recorded on Adak as wide zones of volcanic ash in the stratigraphy. The events occurred at about 8000, 6000 and 4000 BP, with a number of smaller events (narrow zones of ash) at 11 500, 10 000, 2 000 and 1000 BP. Palynological evidence for significant change in the vegetation is however limited, with variations in the quantity of pollen of the Gramineae and Cyperaceae being typical, rather than any obvious extinctions.

The island of Umnak is about 600 km from Adak and has five volcanoes. The stratigraphy of this site shows four major levels of volcanic ash at about 9500, 9000, 7000 and 3000 BP. None of the dates coincide with Adak, suggesting local volcanic events. On two occasions, at 9000 and 3000 BP, the ash layers coincide with significant changes in vegetation. At 9000 BP there is an increase in *Salix* pollen and at 3000 BP there is a sudden decrease in *Salix* pollen. These changes in *Salix* may, however, coincide with changes in climate, as suggested by Heusser (1973), and be uncorrelated or only partially correlated with volcanic eruptions.

The island of Unga has no volcanoes but is about 70 km from Pavlof volcano (2526 m) on the Alaskan peninsula. The cores from Unga and the

Alaskan peninsula (55 km from Pavlof), not surprisingly, have ash layers of comparable ages at about 9500, 9200 and 4000 BP. In these cases it is clear that significant changes in vegetation resulted from these eruptions and, as for Umnak, the occurrence of *Salix* appears to be stimulated by the volcanic event at around 9500 BP. On the Alaskan peninsula the eruptions at about 9500 and 4000 BP were also closely followed by an increase in *Betula*, a species requiring open ground for seed germination and establishment. The levels of both *Empetrum* and *Equisetum* also appear to be sensitive to the events following an eruption. Volcanic disturbance can therefore exert a strong influence on the local vegetation but not necessarily universally. The importance of these acts of disturbance are neatly described by Heusser (1983): 'The effect of late-Quaternary eruptive fallout on the vegetation was apparently differential, creating a patchwork of communities, many of which became seral and acted to offset influences resulting from climatic change and from autogenic processes acting at the community level'.

Unfortunately the palaeoecological evidence from the Aleutian Islands provides little information on the time course of the migrations shown on Fig. 5.7. None of these species can be identified by pollen and cannot therefore be differentiated from other, wide ranging species. Some species such as *Betula* and *Alnus* probably established on the Alaskan peninsula and Unga at about 6000 BP, or earlier (Heusser, 1983), and again became important at about 2000 BP. In neither instance did the species spread westward into the Aleutians. Migration was obviously inhibited by features which were probably climatic but not volcanic.

On the other hand, the rise in the abundance of spores of the Polypodiaceae was more or less synchronous on the Alaskan peninsula, Umnak and Adak at about 10 000 BP, implying very rapid rates of migration.

It is unfortunate that there is such limited historical insight into these species of restricted distribution; however these observations on the past and present vegetation of the migratory stepping stones of the Aleutian and Commander Islands provide a unique example for studying the process of migration and its sensitivity to climate, disturbance and habitat diversity. In the context of the climatic control of the distribution of taxa, the investigation has provided a number of important features which must be considered when attempting to explain geographical patterns of plant distribution. The probability of dispersal declines geometrically from the initial source of propagules, irrespective of the actual mode of dispersal between the islands, although the precise pattern of distribution is also strongly dependent on the available area for colonisation, and the probability of disturbance. The observations on Attu show that an area of increased diversity can arise, perhaps with a climatic cause (the island is lower lying than the others considered here), or perhaps as a result of a low level of disturbance. C.J. Heusser (personal communication), however, has found layers of volcanic ash in a core from Attu whether the ash is from a local or distant source is unknown. Another feature of the history of Attu is its late escape from glaciation at about 7000 BP, compared with about 10 000 BP at the other sites. Historical features cannot clearly provide any obvious clue to the diversity of Attu.

The probability of dispersal may ultimately control the colonisation of areas which are nevertheless suitable for survival, such as the islands of Umnak, Unalaska and Unimak. Carter & Prince (1981) have provided an elegant explanation for just this phenomenon in terms of epidemic theory for short lived herbaceous species.

Hulten (1960) and Tindall (1979) describe observations which might be interpreted as testing such an hypothesis. The nearest native sites to the Aleutian Islands of *Picea sitchensis* are over 1000 km distant.The species has, however, survived on the island of Unalaska since planting in 1805. In 1979 *Picea* was firmly established, had formed a stand of trees some 15 m high and was actively regenerating from seed.

Vertical diversity

An analysis of the ecology of the Aleutian Islands at the present and in the past has been most productive in providing a means of investigating dispersal and migration, the effect of dispersal on plant distribution and the importance of climate on the processes of dispersal and migration. The Aleutian Islands are arranged as a line of 'stepping-stones' and therefore present a much simpler view of species dispersal and migration than the areal process over continents. Nevertheless, the earlier discussions on the palaeoecological evidence for migration show that species tend to move as linear fronts, very much akin to the observations on the Aleutian Islands.

Islands may also prove to be useful starting points for other considerations on the impact of climate on plant distribution. One such consideration arises from a comparison of the vegetation of a selection of islands differing markedly in latitude and climate, but of rather similar physiography. It is feasible to provide a very wide, if not all embracing, survey of the vegetational types or plant communities on one island.

The constitution of the communities will be dependent on many small-scale features such as altitude and soil type and, in addition, on the dispersal of species to the island and the ability of these species to grow in the overall climate of the island. So an island which is in the range of latitudes where the characteristic continental vegetation is tundra, for example, because of a mixture of snow cover, low winter temperatures and a low heat sum in the growing season (Chapter 4), would also be expected to develop tundra vegetation. Seeds of temperate or tropical species which may disperse to the island will not survive because of the adverse climate.

Classification of the major vegetation types on the island could therefore provide a quantitative measure of the similarity and dissimilarity between these communities. The previous chapter emphasised the importance of low temperatures in controlling the distribution of different physiognomic types. The most complex vegetation type includes the rain forests of the tropical and equatorial regions. Less complex are the broad-leaved deciduous forests and coniferous forests in the temperate and boreal regions, with the most simple zone to be found in the tundra. Structural complexity in this sense is a measure of the vertical zonation and diversity of species from the ground level to the canopy crown.

Now if a number of islands are selected with a similar range of habitat diversity, does it follow from the preceding discussion that climate will control structural complexity so that different communities will be more similar or more dissimilar with changes in latitude? In other words, how important is climate in controlling vertical and horizontal diversity? This question has been investigated by applying a Bray and Curtis ordination (Bray & Curtis, 1957), following the advice of Gauch (1982), to comprehensive vegetational analyses of the vascular plants of three islands at different latitudes. The most northerly island to be selected was Jan Mayen Island, at 71° N. The area of the island is 380 km^2 and rises to 2277 m with the glaciated volcano of Beerenberg. Lid (1964) is the source of quadrat data (stands) for the island.

The second island was Gough Island at 40° S. The area of the island is smaller than Jan Mayen at about 75 km^2 and rises to an altitude of 910 m at Edinburgh Peak. Wace (1961) has provided a quantitative description of the vegetation of the island.

The last island to be selected was Santa Cruz, in the Galapagos Islands at a latitude of 0° S. Santa Cruz has an area of 986 km^2 and rises to an altitude of 864 m. Hamann (1981) has made extensive vegetational studies of the Galapagos Islands and is the source of quantitative information of the major vegetational types. Unfortunately, comparable data on the vegetation of the Galapagos Islands with areas similar to either Gough or Jan Mayen were too limited to be of any value in this context.

Two dimensional ordinations are presented in Figs. 5.8, 5.9 and 5.10 for the major communities on the islands, with the axes providing a

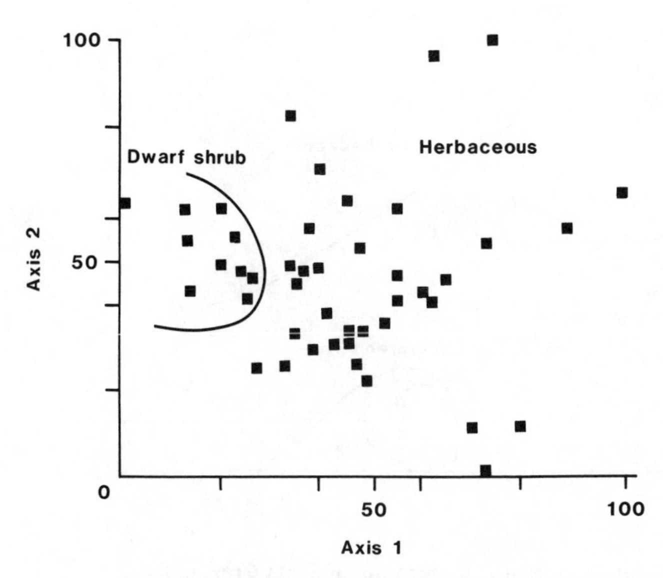

Fig. 5.8. Two-dimensional ordination of the vascular plant vegetation of Jan Mayen Island, 71° N.

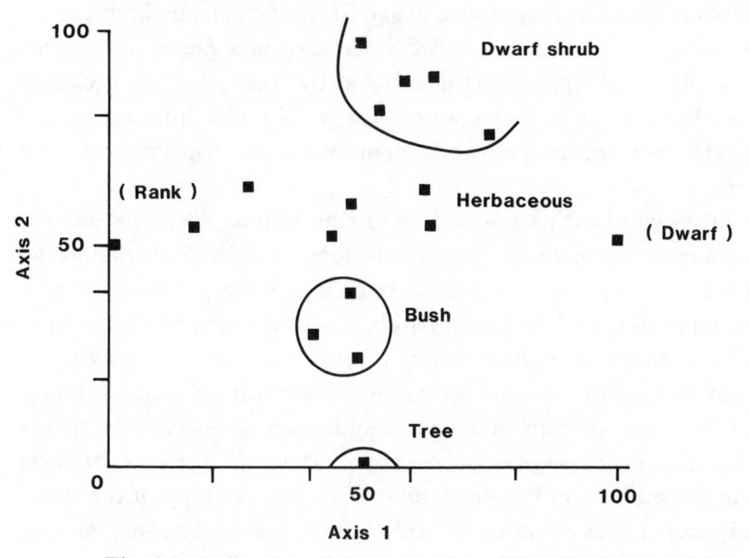

Fig. 5.9. Ordination of the vegetation of Gough Island, 40° S.

measure of the similarities in species composition between the different stands: the smaller the difference on the axes, the greater the similarity between the stands. Contours have been drawn on the ordinations in order to delimit the stands in terms of growth form, defined in terms of increasing height, as herbaceous, shrub, bush and tree.

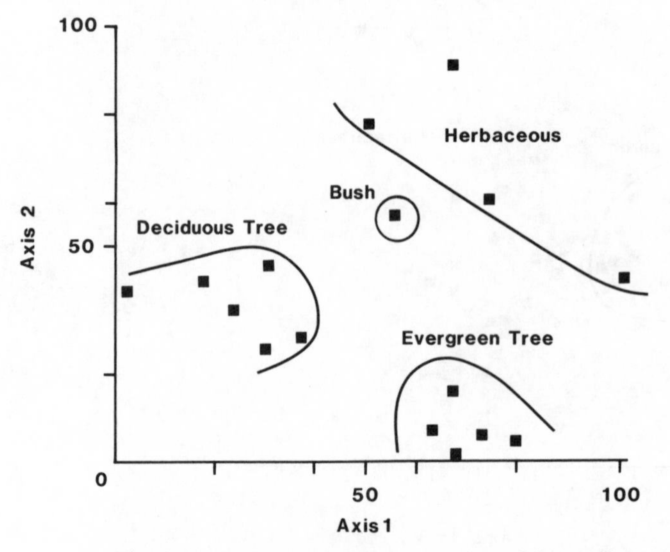

Fig. 5.10. Ordination of the vegetation of Santa Cruz, 0° S.

Only two growth forms are found in Jan Mayen: shrub and herbaceous. The two major dwarf shrubs are *Salix herbacea* and *Empetrum hermaphroditum*. Although the stands dominated by these shrubs are closely related in the ordination, they are not greatly dissimilar from the stands dominated by herbaceous species which, in some cases, also included *Salix herbacea*.

All four growth forms are found on Gough Island, albeit sparsely for the tree, *Sophora macnabiana*. The stands dominated by *Sophora* and the bush *Phylica arborea* were three-layered, or storeyed, consisting of a uniform upper storey of the tree or shrub, a central storey mainly of ferns and a lower storey, or ground cover, of both ferns and bryophytes. A number of the species from the central and lower storeys were not found in open sites, evidently with an obligate requirement for shade. The stands of the dwarf shrub *Empetrum rubrum* are separated from the herbaceous species in the ordination but are similar in that the canopy of the stand included some species common to both stand types, in the same manner as on Jan Mayen Island.

The stands on Santa Cruz are clearly separated by the ordination into stands of deciduous trees, evergreen trees and herbaceous species (only one stand of scrub was published). The deciduous forests consisted of a diverse upper canopy of 15 or more species such as *Bursera graveolens*, *Zanthoxylum fagara*, *Opuntia scouleri* and *Jasminocereus thouarsii*. Hamann (1981) also recorded a shrub and a herb layer beneath the canopy,

Table 5.2 *Island climates*

Island	Latitude	Temperature (°C)		Total annual precipitation (mm)
		Annual mean	Absolute minimum	
Santa Cruz	0 ° S	23.8	17.8	364
Gough	40 ° S	11.7	2.0	3225
Jan Mayen	71 ° N	− 0.2	−27.9	628

and presumably a ground layer is also present. The evergreen forests had the same vertical arrangement of four strata (assuming the presence of a ground layer) but differed in the dominant species of the canopy, which included *Scalesia pedunculata* and *Psidium galapageium*. There is rather little overlap in species between the evergreen and deciduous forests, and indeed with the stands dominated by herbaceous species or evergreen species of bush. The dominant species of evergreen bush or scrub were *Miconia robinsoniana* and *Psychotria rufipes*. This stand, again assuming the presence of a ground layer, consisted of only three strata, with two strata in the herbaceous stands.

The climates of these three islands are quite distinct, as shown on Table 5.2, with data from Wace (1961), Hamann (1979) and Müller (1982). Both the mean and minimum temperatures decline with latitude, although the islands do not fall into a monotonic series with respect to precipitation. Santa Cruz is considerably drier than the other islands, particularly so when the high temperatures are also considered. Not surprisingly, therefore, local variations in the hydrological budget appear to control the distribution of the mesophytic evergreen forest and the xerophytic deciduous forest, which is in keeping with such considerations described in Chapter 4.

Small-scale variations in local topography and hydrology are typical features of mountainous islands such as Gough, Jan Mayen and Santa Cruz. These variations clearly lead to local variations in the distribution of species, but within the constraints of the general climate.

In the very wet climate of Gough Island, 40% of the vascular plant species are understorey pteridophytes, which are generally (but certainly not entirely) considered to be drought intolerant (Richards & Evans, 1972; Page, 1979*a*, *b*). On the much drier island of Santa Cruz, pteridophytes account for about 16% of the vascular plants (van der Werff, 1983). On Jan Mayen Island which is wet but cold, the pteridophytes only account for 5% of the vascular plants, although one species, *Equisetum arvense*, is particularly widespread and often locally dominant.

These examples of the occurrences of pteridophytes highlight one of the many problems in phytogeography, which is the large swing between geographical areas in those features of climate which may be critical in controlling the distribution of plants. On Gough and Santa Cruz it can be suggested and tested experimentally that water availability is critical, whilst on Jan Mayen it appears that low temperatures may be more important.

One other critical feature is the correlation, at least, between the increase in vertical diversity and temperature. The increased diversity created by the locally dominant, drought and high irradiance-insensitive trees, ensures the occurrence of a cool and moist shade, which may be an obligate habitat for the survival of many pteridophytes.

Elsewhere, under a coniferous canopy, Young & Smith (1979) have shown that the short periods of high irradiance and leaf temperature during sunflecks can exert a strong influence on the local distribution of the ground layer species, *Arnica cordifolia* and *A. latifolia*. For these species it was found that *A. latifolia* was more common where the shade environment was most constant with infrequent sunflecks, whilst the reverse was true for *A. cordifolia*.

It appears therefore that the climatic amelioration associated with an equatorial shift in latitude leads to an increase in vertical and horizontal diversity, with greater dissimilarity between stands. This climatic control of vertical diversity is also critical in increasing the local range of habitat diversity. In this respect, Jan Mayen Island is most restricted, with limited vertical diversity.

The impact of variations in life-cycle characteristics on plant distribution

These considerations of vertical diversity, and in particular the distribution of pteridophytes, suggest the need for more detailed investigations on the life-cycles of species, to determine the importance of specific variations in the life-cycle in limiting the distributions of plants to deep shade, full sunlight or xerophytic or mesophytic habitats. It follows from Chapter 4 that the relatively large-scale mesoclimate (Woodward & Sheehy, 1983) of an area is critical in controlling the dominant physiognomy. Within such an area there may be abundant smaller-scale variation in the distribution of species. Some of this variation may be because of vertical diversity but local variations in the dominance of canopy species, for example, can not be so. The control of this variation might be accounted for by random processes, such as dispersal on the one hand or, on the other, in a more deterministic manner by subtle and specific variations in life-cycle characteristics. The investigation of the Aleutian Islands has shown that both random and deterministic controls of distribution occur in tandem. The remainder of this chapter will be concerned with an analysis of this determinism, viewed as subtle differences between species in life-cycle characteristics.

Harper & White (1974), Grubb (1977) and Harper (1977) have emphasised the importance of variations at all stages in the life-cycle in controlling plant distribution. Whereas in Chapter 4 the emphasis was on the adult phase, in this instance of local variations in distribution, it is clear that all stages of the life-cycle must be considered.

A simple, initial example is provided by *Sophora macnabiana*, which is the only tree found on Gough Island. Wace & Dickson (1965) found that seeds of this species (and a number of other species on Gough) are able to germinate after at least three years of immersion in sea water. Of the available sources of seeds that may reach the island by sea, only those which are tolerant of long periods of immersion, and which are appreciable in number, can colonise the island. As far as this colonisation is concerned, the dispersal phase of the life-cycle is most crucial. The end result of the successful dispersal and establishment of *S. macnabiana* is the development of a vertical diversity of species which is associated with *S. macnabiana*. The mechanism by which this diversity develops, presumably with the exclusion of some species, can be studied experimentally.

Such experimental studies have been attempted in herbaceous vegetation and it is possible to find a range of experiments at different latitudes which demonstrate the effect of broad changes in climate on the local abundance of species. A selection of such results is shown on Fig. 5.11, with species abundance measured as areal extent (plant cover). Figure 5.11(*a*) (from Chapin & Chapin, 1980) shows the response of a cleared area of tundra in Alaska (65° N) to seeding by an exotic species, *Lolium perenne*, itself not found naturally beyond about 55° N in North America. Before clearance (years 0 to 1) the area was completely vegetated with tundra species, most frequent of which were *Carex bigelowii* and *Eriophorum vaginatum*. The area was cleared of vegetation and sown with the temperate species, *L. perenne*. At the end of the first growing season the cover of *L. perenne* had reached 20%. The cover of *L. perenne* subsequently declined to 2% at the end of the third year and became extinct by the fifth year. During this time *C. bigelowii* and *E. vaginatum* invaded the cleared area and increased in abundance, forming a nearly complete cover by the fifth year. In this case, the extinction of *L. perenne* is unlikely to result from

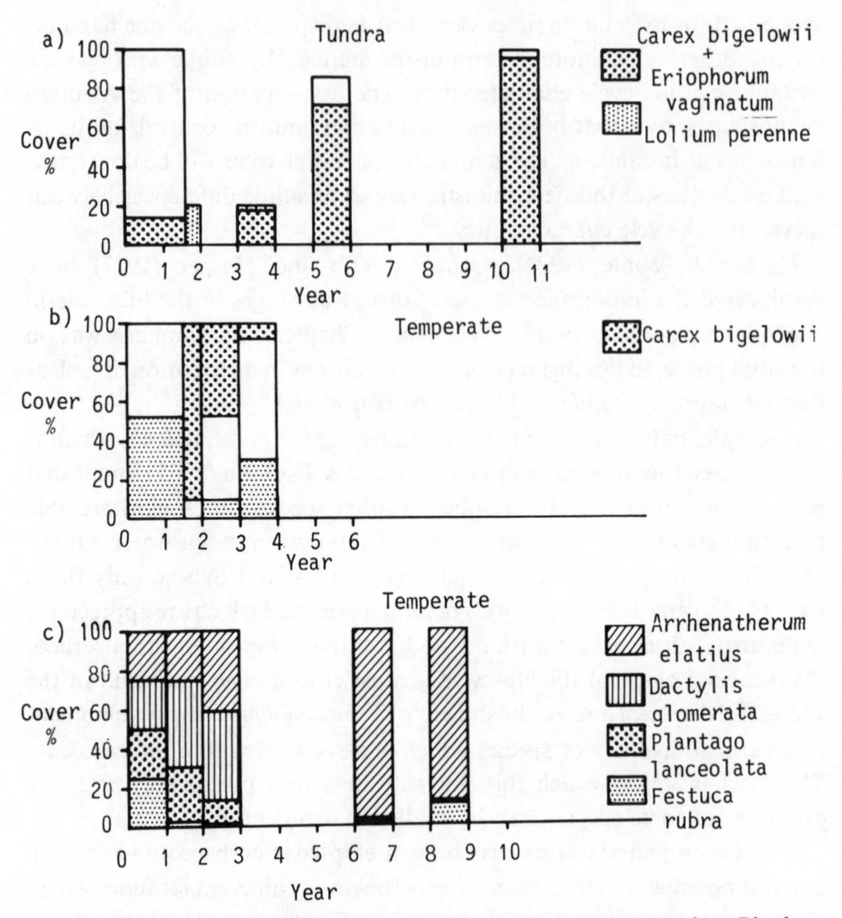

Fig. 5.11. Growth of transplant species in natural vegetation. Blank parts of the histogram in (*a*) and (*b*) represent no data.

competitive relationships with the native species because the cover of these species was quite low, suggesting that the extinction of *L. perenne* is more likely to be caused by its inability to survive the local climate. It is also interesting to note that the native species took at least 5 years to recolonise this cleared area in the cold tundra, emphasising the limited potential for growth in this climate, as discussed in Chapter 4.

Figure 5.11(*b*) shows a similar type of experiment in the British Isles (F.I. Woodward, unpublished) at a latitude of 54° N and an altitude of 90 m. In this case the local vegetation included *L. perenne*, which was locally dominant. Small areas of this vegetation (0.06 m^2) were cleared in the early spring (year 1) and plants of *C. bigelowii*, which does not occur naturally below about 600 m in England, were introduced, providing a cover of about 90%. The cover of *C. bigelowii* declined to 50% by the end of the second year, following invasion and shading by the local species, including *L. perenne*. *C. bigelowii* had all but disappeared by the end of the third year. In this case, unlike the tundra, it appears that competitive relationships with local species are critical in controlling distribution, because *C. bigelowii* grows well and completes its life-cycle at low altitudes on the same soil but in the absence of interspecific competition.

Figure 5.11(*c*) describes a long-term survey of the growth of four herbaceous species in a mixture. In this case all of the species occurred locally. This experiment, by Grubb (1982), was established in Cambridge and initiated by clearing areas of vegetation, followed by sowing with equal quantities of seeds of *Arrhenatherum elatius*, *Dactylis glomerata*, *Plantago lanceolata* and *Festuca rubra*. The fates of the four species were traced for 9 years. By the sixth year the rank grass, *A. elatius*, had established its dominance over the community, while *D. glomerata* and *P. lanceolata* had been eliminated, presumably through competition with *A. elatius*. However the small grass *F. rubra* persisted in the community, even increasing its cover by the eighth year. *F. rubra* remained established as an understorey component in deep shade but failed to reproduce sexually.

Survival of *F. rubra* but the extinction of *C. bigelowii* in temperate grassland (Fig. 5.11(*b*)) clearly relates to specific differences in shade tolerance of the adult phase of the life-cycle. The mechanism of exclusion or survival is not known. However, in a similar problem, Turitzin (1978) has developed a model based on the photosynthetic and architectural features of plants to show that these characteristics are sufficient to explain the occurrence or absence of the grass *Bromus mollis* in various competitive situations. These cases of rather subtle variations in life-cycle characteristics are quite critical in influencing local distribution in a climatic envelope which is within the tolerance of the species.

A review of investigations on the climatic control of plant distribution shows that the causes of distribution limits are generally considered, or have usually been found, to be due to one limiting phase of the life-cycle. In their classic studies on the small-leaved lime, *Tilia cordata*, Pigott & Huntley (1978, 1980, 1981) developed, tested and verified the hypothesis that the temperature sensitivity of pollen tube growth, and a short period of stigmoid and stylar receptivity, could together account for the northern limit of regeneration of the species in the British Isles. Pollen tube growth was insufficiently rapid to allow fertilisation to take place by the time that the styles aborted. In Finland, however, pollen tube growth and fertilisation are successfully completed in the warmer, continental climate, but the temperature falls so rapidly during late summer that embryo and endosperm development are arrested and incomplete. The result, as in the British Isles, is that there is little or no regeneration from seed (Pigott, 1981), although the cause is clearly dependent on the climatic differences between the maritime British Isles, with a rather small annual amplitude of temperature, and continental Finland, with a large annual amplitude of temperature.

In terms of the ideas of response times developed in Chapter 2, it seems strange that a short-term process such as this could account for the limits to the geographical spread of the species, even though geographical spread has a considerably longer response time. For *T. cordata*, it emerges that a threshold phenomenon is the critical factor. This phenomenon is shown on Fig. 5.12 (6), where it may be seen that pollen tube extension fails to occur when the temperature falls to or below 15 °C. This temperature is critical for pollen tube growth but not critical for plant survival. Should a long term cycle in climate, with a cycle time in years or millennia (Chapter 2), lead to a slow but gradual increase in temperature, then it may become possible for range extension, assuming that adults have survived the sterile interlude. Indeed it is likely that the present limit of *T. cordata* was determined in the warmer climate of about 5000 BP (Pigott, 1981). Presumably the climatic limit to pollen tube extension may have occurred at more northerly sites at that time. There are other published examples which have demonstrated the critical importance of a particular stage in the life-cycle to plant distribution, e.g. Mooney & Billings (1961), Woodward (1975), Woodward & Pigott (1975), Bell & Bliss (1979) and Black & Bliss (1980). The case of *T. cordata* describes a non-lethal threshold effect. Lethal threshold effects may also be observed and are clearly in keeping with the hypotheses generated in the previous chapter.

Yaqub (1981) investigated the altitudinal distribution of two perennial herbaceous species, *Eupatorium cannabinum*, which has an upper limit of about 270 m in Wales, and *Verbena officinalis*, with an upper limit of 220 m.

V. officinalis overwinters as an evergreen rosette, whilst *E. cannabinum* is winter deciduous. Mature individuals of both species were transplanted in 1977 to a range of altitudes up to 400 m, above the natural limits of both species. The winter of 1977 and 1978 was cold, but with little snow, and the evergreen rosettes of *V. officinalis* died when the minimum temperature reached −7.8 °C. *E. cannabinum* survived at all sites, showing an insensitivity to the local winter climate. In that particular year, the extinction of *V. officinalis* coincided with the observed altitudinal limit.

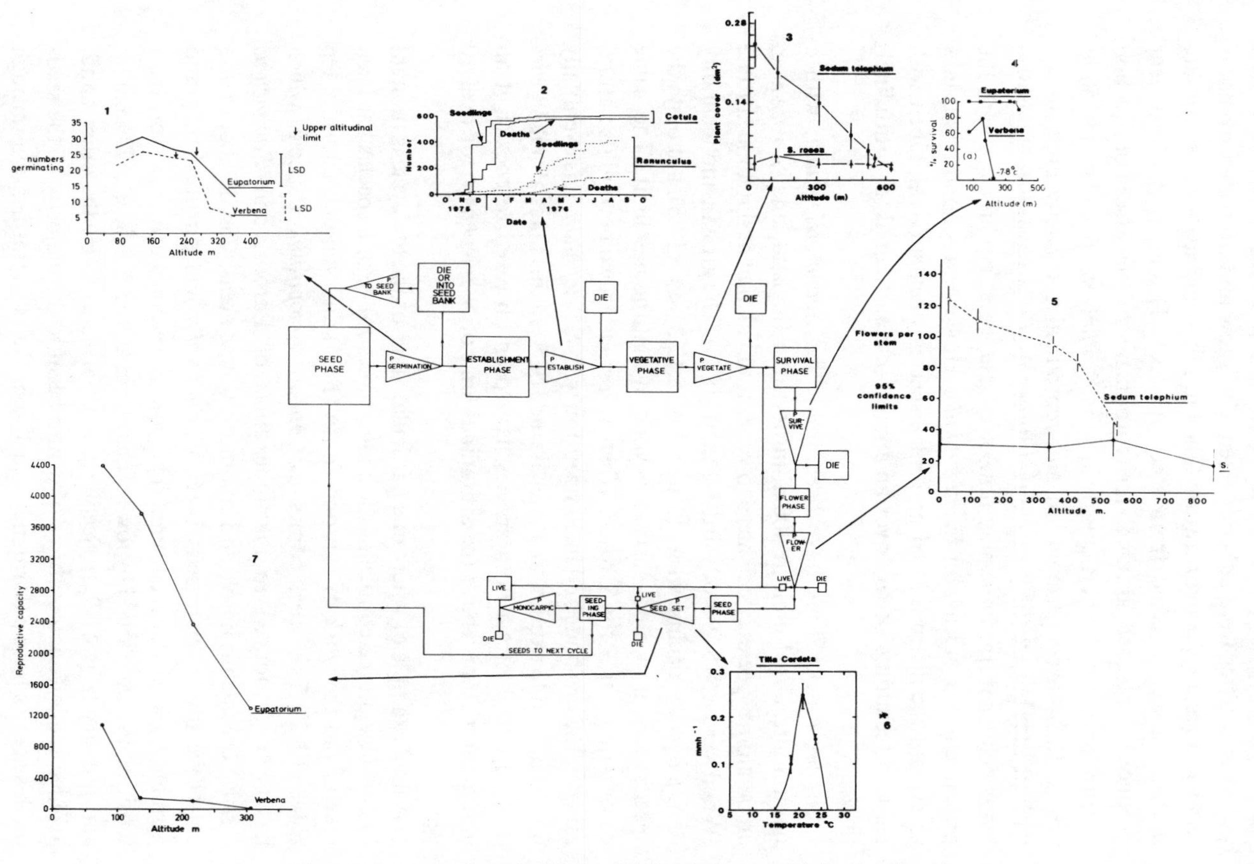

Fig. 5.12. Examples of differential life-cycle sensitivity to climate.

However, when the experiment was repeated in the two successive winters, survival of *V. officinalis* was observed well above the natural limit. In these years the rosettes were either covered with snow and protected from low temperatures, or the temperature failed to fall as low as −7.8 °C. In fact both species established from seed (Fig. 5.12,1) and in the next year produced viable seed, although the reproductive capacity (seed production × germinability) was particularly low for *V. officinalis* above about 150 m (Fig. 5.12,7). Seed germination and reproductive capacity are reduced with altitude but are still positive at altitudes above the natural geographical limits. This natural limit may not, of course, be at equilibrium with the present climate, at least for *E. cannabinum*. However it is not obvious how to explain the distributional limit of a species when survival, albeit with reduced fecundity, occurs over an appreciable altitudinal (or latitudinal) range.

Similar observations have been found for *Sedum telephium*, with a natural altitudinal limit at 400 m in northern England, and *S. rosea*, an arctic-alpine species which has a lower altitudinal limit at the same altitude (Woodward, 1975; Woodward & Pigott, 1975). The growth and fecundity of *S. telephium* decline with altitude (Fig. 5.12,3 and 5.12,5), although the species can grow and survive beyond its altitudinal limit. The same observation is true for *S. rosea*. When *S. rosea* is grown beyond its natural range in the lowlands and in competition with *S. telephium*, it is eventually out-competed to extinction. At an altitude of 50 m this takes about 6 years to occur. In contrast both species have been observed to coexist at an altitude of 550 m, above the altitudinal limit of *S. telephium*, for over 10 years.

A final example of where one particular stage of the life-cycle is thought to be crucial in the control of the geographical range of a species has been taken from the work of van der Toorn (1980) and van der Toorn & ten Hove (1982). This work compares *Cotula coronopifolia*, a species which has been introduced to maritime areas of Europe, and *Ranunculus sceleratus*, which is native and common in these areas. Van der Toorn and ten Hove found that seeds of *C. coronopifolia* had no dormancy and germinated in the late autumn. The young seedlings are, however, frost sensitive and were killed by low temperatures at about −8 °C in December and January (Fig. 5.12,2). Seedlings of *R. sceleratus*, on the other hand, avoided the low temperatures of winter because the majority of the seeds on this species have a dormancy which is broken by stratification through the winter. As a result, germination occurs in the spring, assuring a moderate survival of seedlings by avoiding winter frosts (Fig. 5.12,2). These particular observations, however, belie reality in that *C. corono-*

pifolia is, on occasion, able to maintain and consequently extend its natural range.

For some of the species shown on Fig. 5.12 one phase of the life-cycle is critical for survival. However, some species are able to complete all phases of the life-cycle even beyond an observed limit, but with little apparent effect on the geographical range.

The observation that no particular stage of the life-cycle is critical in controlling distribution, but that progressive reductions in probabilities of survival or production at a number of stages are more likely causes, e.g. *E. cannabinum*, *S. telephium* on Fig. 5.12, are also possible and have indeed been clearly established by Salisbury (1926, 1932), Dahl (1951) and Davison (1977). A diagrammatic interpretation of this view may be seen in the centre of Fig. 5.12, which describes a generalised life-cycle in which the relative number of individuals of each stage or phase is represented by the size of each box. The probability that individuals at a particular stage will reach the next stage is indicated by the triangles, pointing in the direction of the cycle. As expected, the number of individuals at each stage declines through the life-cycle because of mortality. Harper (1977) provides many examples of such a change.

Given this description of a continuous life-cycle, it is possible to interpret the natural geographical limit of a species as the sites at which insufficient numbers circulate round the cycle to maintain an equilibrium or expanding population. In this instance extinction may be progressive, taking a number of years to occur, perhaps interspersed with more and less productive years. In this case there is no threshold effect and indeed this progressive decline may be the most likely demise of a species at its geographical limits, i.e. not immediate but progressive extinction, perhaps in combination with reduced competitive ability.

Woodward & Jones (1984) predicted that such a response should occur for the lowland species *Potentilla reptans*, at a site about 240 m above its observed altitudinal limit (610 m on Fig. 5.13(*b*)) in the British Isles. They also observed the number of individuals of a range of species which survived from one stage of the life-cycle to the next in 90 mm diameter patches cleared of the native vegetation. In addition they established the density dependence of each probability in the life-cycle, by sowing the patches at a range of densities. The resulting observations on each stage of the life-cycle over 2 years, and for a range of densities, species and altitudes, are too extensive to be simply used for predicting survival. However the actual probabilities of the stages shown in Fig. 5.12 can be inserted into a (mathematical) transition matrix, which when multiplied by the number of individuals at different stages, or phases, serves to predict the growth in number of the whole population. Leslie (1945), Lefkovitch (1965) and Usher (1969, 1972) have described the mathematical details of the technique. The technique has recently proved to be valuable and applicable to a range of vegetation types, such as annuals (Law, Bradshaw & Putwain, 1977), herbaceous perennials (Sarukhan & Gadgil, 1974; Woodward & Jones, 1984), temperate trees (Enright & Ogden, 1979) and tropical trees (Hartshorn, 1975; Pinero, Martinez-Ramos & Sarukhan, 1984).

Continued post-multiplication of the transition matrix of probabilities by the column vector of population density at each stage leads to a deterministic model of population growth, described by:

$$N = N_0 e^{rt}, \qquad \textbf{(2)}$$

where N is the population number at time t, with N_0 the initial population size and r the intrinsic rate of increase in number. When r has a value of zero, population increase is also zero and N is constant with time, reflecting a perfect balance of birth and death. When r is greater than zero, the population will be increasing in number, and the population will decrease when r is less than zero.

Repeated multiplication of the model with $r > 0$ predicts continuous but unrealistic exponential growth, although a stable population in terms of the relative numbers in each stage or phase is soon established.

Active growth of natural populations ultimately leads to a density dependent effect (usually reduction) on one or more of the probabilities in the matrix. Law (1975) and Woodward & Jones (1984) have shown that the effect of this density dependence is an asymptotic trend in population growth, reaching a more or less steady population size.

If **(2)** is differentiated, then the growth rate in a continuously exponential phase is:

$$\frac{dN}{dt} = r\,N\,. \qquad \textbf{(3)}$$

However, the growth rate falls to zero when the population size reaches an asymptote, so that if the population size at the asymptote is K, then:

$$\frac{dN}{dt} = r\,\frac{N(K-N)}{K}. \qquad \textbf{(4)}$$

This equation is known as the logistic equation and K is a measure of the carrying capacity of the local environment. The bracketed part of the equation varies from 1 (when $N = 0$) to 0 when $N = K$, and simply modifies **(3)**, to account for the carrying capacity.

The application of density dependent transition matrices to predicting

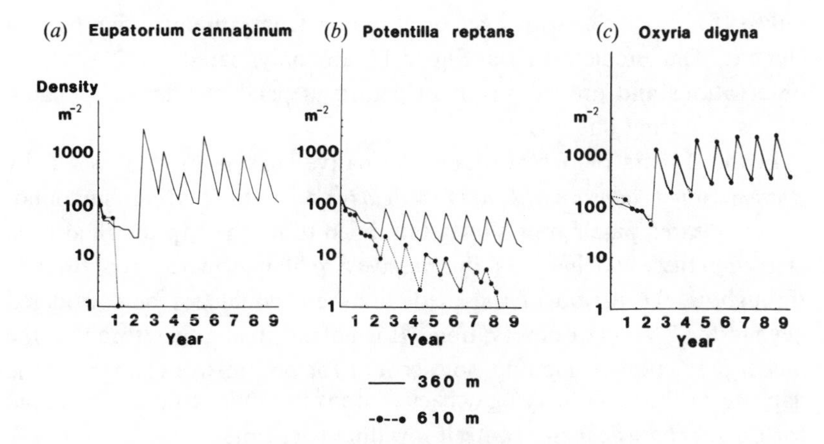

Fig. 5.13. Predicted population growth of three species.

population growth is shown in Fig. 5.13. Three predictions based on field observations are shown for (*a*) *Eupatorium cannabinum* growing at an altitude of 360 m (the populations died at 610 m); (*b*) *Potentilla reptans* and (*c*) *Oxyria digyna*, growing at 360 m and 610 m. Symonides (1979) observed very similar patterns of population increase for *Corynephorus canescens* when it invades sand dunes. In both cases the initial increase in population density was rapid, then moving to a more or less constant equilibrium density. Annual cycles of population density, in the main representing the births and deaths of seedlings, were superimposed on the longer term increase in density.

The matrix approach to predicting population growth, resulting from variations both subtle and obvious of different stages of the life-cycle, is particularly valuable. This is best shown by *P. reptans* when grown at 610 m (Fig. 5.13(*b*)). At this altitude the prediction is that the species should decline slowly and ultimately become extinct because of overall reductions in the success of a number of stages of the life-cycle, such as seed germination, winter survival and fecundity. The technique is not required to explain the demise of *E. cannabinum* at 610 m, which was killed outright during its first winter by frosts of $-8°$ C. This type of process, with mortality resulting from extremes of temperature, may also explain the extensive and uniform boundaries to vegetation as seen particularly clearly at the tree line on mountains. Where investigated, the clear cut boundary appears to result from frosts rather than through the results of competition with plants from higher altitude (Wardle, 1985).

The model predictions can fail to mimic reality, however, if the data on probabilities are not updated to account for any significant changes linked, for example, with age, or resulting from annual differences in climate. The predictions on Fig. 5.13 are only based on 2 years of observations and are the result of density dependent and independent processes during this time.

In the 2 years of observations, the native vegetation (dominated by *Anthoxanthum odoratum*, *Calluna vulgaris* and *Festuca ovina*) surrounding the cleared patch grew slowly but failed to fill the gap that had been created in these cool habitats. Progressive gap filling will occur continually throughout the period of the predictions but could not be considered beyond the 2 years of observations. It is not surprising therefore that the model predictions, which do not account for progressive changes in the gap size or the local carrying capacity, are very similar to those observed for *Corynephorus canescens*, itself invading bare sand.

The absence of information on progressive gap filling may indeed account for the inability of Woodward & Jones (1984) to account for the lower altitudinal limit of the arctic-alpine species *Oxyria digyna*. Population growth of this species shows little sensitivity to altitude, either within or outside (below) its natural range. Indeed Salisbury (1926) and Dahl (1951) have pointed out many examples of upland species which can complete their life-cycles in lowland environments far from their natural range. This success may be observed for isolated plants, but in competition with more rapidly growing lowland species extinction appears to be inevitable (Woodward & Jones, 1984; Woodward & Pigott, 1975), as was clearly demonstrated in the field for the upland species *S. rosea*, when growing in competition with the lowland *S. telephium*. The field responses, in a simple competitive design, could also be repeated in controlled environments in which temperature alone was changed to simulate changes in altitude. This example provides a clear case for the basis of the mechanism by which climate can control plant distribution on mountains. The poor growth of *S. rosea* in the lowlands, or in average lowland temperatures (growing season greater than about 10 °C), is selected against when in competition with more rapidly growing species of the lowlands. In cooler conditions (growing season less than about 10 °C) there is no such selection. Indeed Woodward (1983) provides evidence for selection against rapid growth in the typically windy conditions at high altitude. In the lowlands, therefore, it is obviously important to understand the mechanisms which can account for the demise of the upland species.

The very nature of the experiments which are central to the matrix predictions described above is that both intra- and interspecific competition are measured. If, as might be expected, the upland species *O. digyna* will become extinct in the lowlands, then an improved model should attempt such a prediction. It is known that differences in growth rates are critical in the competition for irradiance (Woodward, 1975, Woodward & Pigott, 1975) and also for soil nutrients and water, yet absolute differences in the growth rates of species can not be extracted from the predictions shown on Fig. 5.13. However Jones (1983) clearly demonstrated that the lowland species *E. cannabinum* (Fig. 5.13(*a*)) had a higher growth rate in the lowlands than *O. digyna* (Fig. 5.13(*b*)).

Competition for space

A more precise model for predicting geographical distributions should aim to embody both life-cycle and growth characteristics of the species under investigation. A starting point for this aim may be met by predicting the ultimate fate of a monospecific stand of seedlings which have germinated in a bare patch of ground, encircled by established plants of the same or different species. The model assumes that, at least in this case, range extension may occur by growth and establishment of individuals in gaps denuded of vegetation.

As time progresses, the patch will be reduced in extent by a gradual inward encroachment of the surrounding plants (but see Chapter 6). If the patch has a diameter of 90 mm (as used by Woodward & Jones, 1984) and the surrounding plants infill at equal rates on all radii, then at a rate of infill of 0.1mm d^{-1}, the patch will be filled in 450 d, while at a rate of 1.0 mm d^{-1} the patch will cease to exist after 45 d.

The invading seedlings in the patch may survive in two ways. One way is by survival in the understorey beneath the encroaching plants. The other way is for the species to grow sufficiently rapidly to equal or overtop the encroaching species, therefore effecting, rather than being affected by, shade.

Species capable of incorporation into the understorey must be shade-tolerant, such as *Festuca rubra* in the shade of *Arrhenatherum elatius* (Fig. 5.11). There is abundant evidence to support this proposal in forests. Examples include the long-term survival in the forest understorey by: *Prunus serotina* in North American forests of *Quercus alba* and *Q. macrocarpa* (Auclair & Cottam, 1971); *Tsuga canadensis* under *Pinus strobus* (Henry & Swan, 1974) and also by *Acer saccharum* and *Ostrya virginiana* under *Pinus strobus* (Peet, 1984).

Contrary evidence for the elimination of shade-intolerant species in shade can also be found in forests. Young individuals of the pioneer and shade intolerant *Betula alleghaniensis* fail to survive longer than three years under a canopy of *Acer saccharum*, *Fagus grandiflora* and *Betula alleghaniensis* in North America (Forcier, 1975); also in North America the occurrence of the short-lived and shade-intolerant *Populus grandidentata* declines rapidly under *Acer saccharum* and *A. rubrum* (Spurr & Barnes, 1980). In addition, the spectral quality of the shade cast by the canopy may inhibit seed germination (King, 1975), so that the life-cycle fails to be initiated.

Shade tolerance can therefore define the ability of a species to survive the encroachment of a gap. Within any canopy the frequency of occurrence and longevity of gaps will be critical in maintaining a diverse mixture of shade-intolerant and shade-tolerant species. This is a universal phenomenon of closed vegetation. In forests the average annual creation of gaps is about 1% of the total area in northern boreal forests (Heinselman, 1973), subalpine forests (Kanzaki, 1984), temperate evergreen forests (Naka, 1982), temperate deciduous forests (Runkle, 1982) and tropical forests (Hartshorn, 1978). Hibbs (1982) was able to predict the success of a species in attaining canopy co-dominance, based on measurements of gap size and the rates of gap infill and sapling growth.

The formation of gaps in forests is due in part to catastrophic disturbances such as fire and storms and also to the death of individuals in the mature canopy. The mortality rate can therefore be critical in maintaining community diversity, which in vertical terms is also climatically controlled, as described earlier. Mortality is also a feature of herbaceous canopies, and in controlled monoculture Kira, Ogawa & Sakazaki (1953) demonstrated that mortality lead to fewer and larger individuals. This process of self-thinning is most obvious for a dense group of small individuals, such as the seedlings attempting to establish in a cleared patch in herbaceous vegetation. Yoda *et al.* (1963) described the relationship between the size or weight (w) of an individual and plant density (N) when self-thinning is occurring by a power relationship:

$$w = cN^{-3/2}. \qquad \textbf{(5)}$$

White (1980) has shown that this relationship is true for monocultures of a number of species, both herbs and trees, with a mean value of 6761 gm^{-2} for c and -1.51 for the exponent.

During the phase of self-thinning no species exceeds the relationship with a much greater weight than would be predicted from (**5**), using the mean values provided by White (1980). If a canopy gap, or patch such as used by Woodward & Jones (1984), is invaded by a number of seeds of one species and the germinated seedlings grow sufficiently rapidly before the gap is filled, then these individuals will reach the 'universal' self-thinning line. From this moment in time, both density and weight are defined by (**5**). At large weights and low densities, the exponent may tend to -1, at which point the plants have reached the carrying capacity of the local environment and any increase in size will be at the expense of a directly proportionate diminution in plant density. The point of change from slopes of -1.5 to -1 appears to be species related and may be climatically controlled (White & Harper, 1970; Kays & Harper, 1974). At the other extreme, before self-thinning occurs, plant growth is apparently density-independent, and is not described by (**5**).

The relationship between weight and density described by (**5**), in the absence of any evidence to the contrary, can be accepted as a universal phenomenon for plants in monoculture, with the exponent of -1.5 (or $-3/2$) describing the marriage of a volume of plant, with a dimension of 3, to an area of soil, with a dimension of 2 (Whittington, 1984). This relationship therefore defines the upper boundary line or constraint relating plant weight and density. All weights smaller than the weight predicted from (**5**) are possible, but none greater.

It has been established that growth rate is critical in determining the survival of patch invaders, and that in the first year of the growth of most perennials sexual reproduction of the patch invaders can be discounted. The patch will be encroached at a rate which, initially at least, is dependent on the growth rates of the species in the existing vegetation. It is assumed in this instance that the individuals of the patch invaders which are encroached and shaded by the advancing front of the existing vegetation are eliminated, i.e. they are shade intolerant juveniles. In addition, once self-thinning occurs in the patch, the numbers of individuals will also decline in the manner predicted by (**5**).

The plant responses to be considered are therefore the density-dependent and -independent rates of growth of the patch invaders and the rate of patch infill. Two species have been chosen to develop and describe the application of a model for predicting the outcome of competition between species invading a gap and those existing around the gap. The selected species are the arctic-alpine species *Oxyria digyna* found in the British uplands and the continental species *Lolium perenne* found in the lowlands. Data on the growth of the species have been obtained in Cambridge, in addition to data from Jones (1983) and Woodward & Jones (1984) for *O. digyna* and Kays & Harper (1974) for *L. perenne*. Kays & Harper (1974) do not present climatic data; however *L. perenne* was greenhouse grown between February and July in North Wales. *O. digyna* was grown over a similar period at 360 m in mid-Wales and in Cambridge. It is clear that more precise measurements of climate are necessary; however this does not detract from the derivation of the following model.

It has been assumed that propagules of either *O. digyna* or *L. perenne*

Table 5.3. *Growth attributes of L. perenne* and *O. digyna*

	w_0 (g)	R (d^{-1})	w_T (g)	R_T (d^{-1})
L. perenne	0.0007	0.076	0.0211	0.018
O. digyna	0.0005	0.044	0.0154	0.009

have invaded a patch of vegetation surrounded by a number of individuals of native species. The initial growth rate of the invaders is density independent and may be described by the relationship (reviewed by Evans, 1972):

$$w = w_0 e^{Rt}, \qquad (6)$$

where w is plant weight at time t, with an initial, or starting weight of w_0. The intrinsic rate of increase, or relative growth rate, is R, and is the rate of increase in weight per unit of plant weight.

Once the plant weight and density relationship reaches the self-thinning line,

$$w = cN^{-3/2} \qquad (5)$$

and (6) can be modified as:

$$w = w_T e^{R_T t}, \qquad (7)$$

where w_T is related to the point of inflexion from the density-independent to the density-dependent or self-thinning line, and R_T is the relative growth rate during thinning. Re-arranging (5) in terms of density,

$$N = \left(\frac{c}{w}\right)^{2/3} \qquad (8)$$

allows density to be related to growth during self-thinning as:

$$N = \left(\frac{c}{w_T e^{R_T t}}\right)^{2/3}. \qquad (9)$$

Mean values of the various parameters in (5), (6) and (7) for the two species are shown in Table 5.3. The constant c has a value of 40 000 g m^{-2} for both species. It emerges that the upland species *O. digyna* has lower growth rates than *L. perenne* in a lowland environment.

Density-independent and -dependent growth of *L. perenne* is shown on Fig. 5.14. The initial seedling density was 10 000 m^{-2}, with a weight of 0.0007 g at time zero. Density-independent growth is shown as the vertical

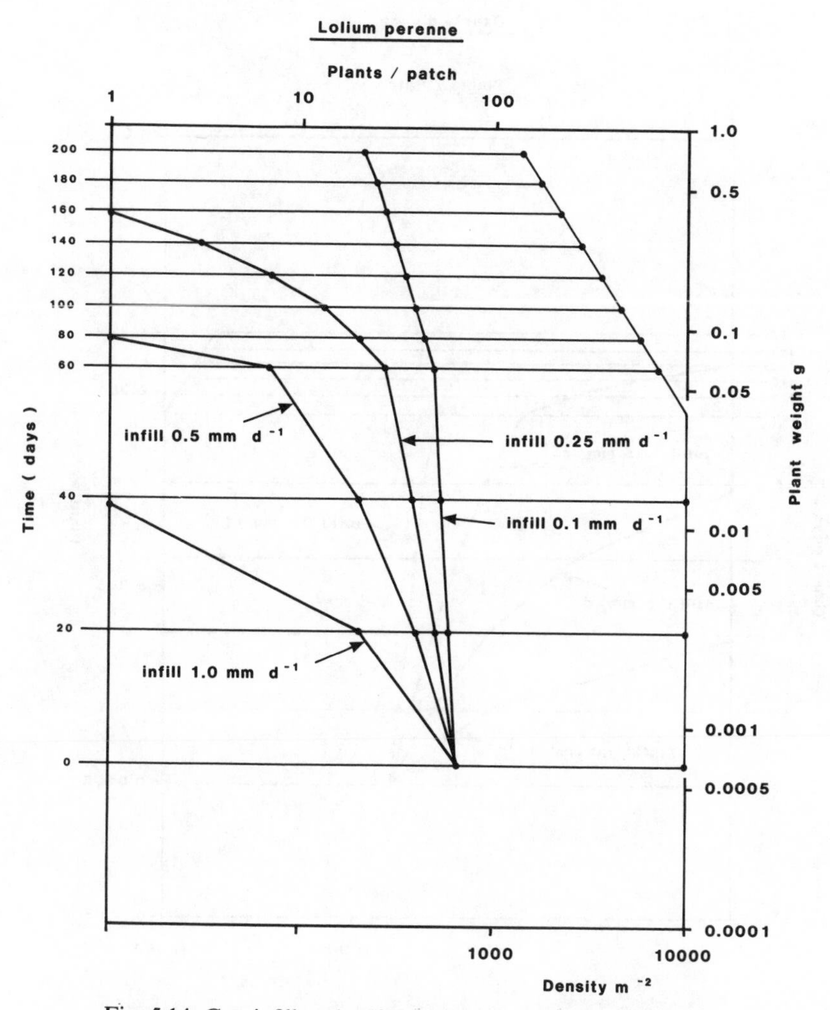

Fig. 5.14. Gap infill and self-thinning of *Lolium perenne.*

line on Fig. 5.14, rising to a point where the plant weight is just less than 0.05 g, at which time subsequent growth in weight is at the expense of density, i.e. self-thinning occurs.

The relative growth rates of the two species have been chosen to be invariant during the two phases of growth. However there should be no problem in introducing both climatically and ontogenetically determined variations in the relative growth rate.

The basic curve relating plant density and weight in Fig. 5.14 is very similar to that presented by Kays & Harper (1974) but with the important addition of a time scale. On the logarithmic scale, the time intervals are

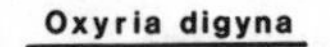

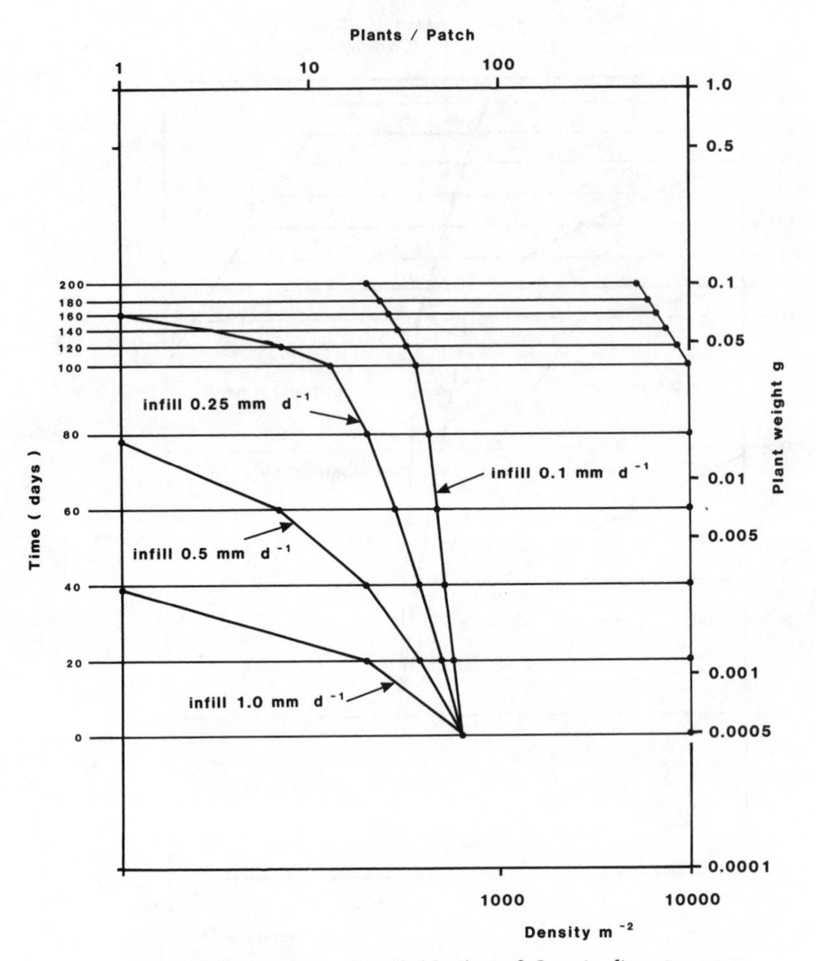

Fig. 5.15. Gap infill and self-thinning of *Oxyria digyna*.

closer together during self-thinning, a direct result of competitive interactions between individuals of the population leading to reduced rates of growth.

The growth curve for *O. digyna* (Fig. 5.15) is similar in detail to *L. perenne* but in this case growth is slower, in both the density-independent and -dependent phases of growth. The growth of *O. digyna* is much less than *L. perenne* and is likely to be less competitive. The significance of these growth curves to the outcome of the struggle for existence in a patch can not be realised without information on the dynamics of gap infill.

The importance of gaps for maintaining species diversity in forests has been a particularly important thrust in plant ecology. Shugart (1984) has recently provided a synthesis of this work which provides an extensive and fascinating model for predicting the effect of gaps on species diversity. It follows from this work that modelling has great value in amalgamating diverse strands of information.

For forest canopies in the temperate zone, published rates of gap infill (radial expansion) range from 70 mm y^{-1} for *Tsuga canadensis* (Runkle, 1982) to 250 mm y^{-1} for *Betula alleghaniensis* [Erdmann, Godman & Oberg (1975) in Hibbs (1982)], and 165 mm y^{-1} for *Quercus rubra* (Trimble & Tryon, 1966). Data from Hibbs (1982) show a mean rate of infill of 98 mm y^{-1} for seven species of tree in the temperate zone. For deciduous species of tree in a temperate climate with cold winters, infill must occur when the trees are in leaf. Assuming an average growing season of 150 d, the average daily rate of infill is 0.65 mm d^{-1}, with a maximum of 1.67 mm d^{-1} for *B. alleghaniensis*, and a minimum of 0.47 mm d^{-1} for *T. canadensis*.

The significance of gaps for diversity in herbaceous vegetation has also been an area of keen interest. The survival to flowering of two biennial species, *Daucus carota* (Holt, 1972) and *Verbascum thapsus* (Gross, 1980), has been shown to be dependent on the rate of gap infill. The survival of either *Lythrum salicaria* or *Epilobium hirsutum* in gaps is related to specific differences in growth rate. If both species invade a gap in the spring then *L. salicaria* has a greater growth rate and excludes *E. hirsutum*. If invasion occurs in the autumn, the reverse occurs because of the greater rate of growth of *E. hirsutum* at low temperatures (Whitehead, 1971). Caswell (1978) has derived a model for predicting the survival of species in gaps based on growth rates and rates of gap infill.

Actual rates of gap infill in herbaceous vegetation appear to be harder to come by than for forests. F.I. Woodward (unpublished results) has measured rates of infill along an altitudinal series in central Scotland. The rates fall more or less monotonically with altitude from 0.7 mm d^{-1} at an altitude of 50 m, to about 0.1 mm d^{-1} at 700 m, in vegetation dominated by a range of species of Gramineae and Cyperaceae. It does appear, therefore, that the rates of gap infill are of a similar order of magnitude for tree and herb canopies, although the rates are rather greater for trees.

The significance of gap or patch infill in the present model is that any individual of the invading species which is encroached has been assumed to die through shading. The numbers of individuals in the patch will therefore decline continuously as the patch is extinguished.

The effect of gap infill alone on a population of invading individuals of

L. perenne and *O. digyna* is shown on Figs 5.14 and 5.15. A circular gap of 90 mm in diameter has been assumed in both cases. A seedling density of 10 000 m^{-2} thus indicates a total population of 64 (plants/patch on the figure) invading individuals. The figures show the impact of four different radial rates of patch infill, from 0.1 mm d^{-1} to 1.0 mm d^{-1}, on population size.

In the case of *L. perenne*, the population is reduced to one individual after 40 d with a rate of infill of 1.0 mm d^{-1}. A solitary individual also remains after 80 d with an infill rate of 0.5 mm d^{-1} and 160 d at 0.25 mm d^{-1}. The weight of this individual ranges from 0.014 g with an infill of 1.0 mm d^{-1} to 0.088g with an infill of 0.5 mm d^{-1} and 0.36 g with an infill of 0.25 mm d^{-1}. At the same stages *O. digyna* is considerably smaller, reaching plant weights of 0.0029 g (1.0 mm d^{-1} infill), 0.016 g (0.5 mm d^{-1}) and 0.069 g (0.25 mm d^{-1}).

If any of the measures of plant size exceed those of the average individual of the existing vegetation, then, as a first approximation, it can be predicted that there is a high probability that this individual will survive. This assumes that the plant is equal or greater in height than the existing vegetation and that it is able to achieve adequate uptake of soil nutrients and water, all critical features which could be included in the model. At this point in time it is expected that the rate of canopy infill should be reduced, dependent on the shade tolerance of the surrounding species.

The operation of the model can be demonstrated by an example. If the mean size of an individual native is 0.08 g, then *L. perenne* would reach this size after about 80 d, and *O. digyna* after a much lengthier period of 180 d. At this stage, self-thinning will have reduced the population size of both species by 37%.

For *L. perenne*, an infill rate of 0.5 mm d^{-1} would reduce the population to about 1 individual, therefore when self-thinning is also included it is clear that *L. perenne* would be excluded. A rate of infill of 0.25 mm d^{-1} would reduce the population to 22 individuals, less 8 individuals (37%) by self-thinning, leading to a total of 14. *L. perenne* should therefore have a high probability of establishment with rates of patch infill of 0.25 mm d^{-1} and less.

For *O. digyna*, the only case where a finite population remains is with a rate of infill of 0.1 mm d^{-1}. In this case the population has been reduced to 24 by infill, less 9 individuals by self-thinning, leading to a total of 15.

This example of two species differing in rates of growth demonstrates that the slow-growing *O. digyna* should become extinct in gaps surrounded by aggressive and fast-growing individuals. *L. perenne*, on the other hand, has a greater probability of survival in the same situation. The application of this model, which relates plant growth to population density, is a logical addition to the matrix model described earlier and would be strongly dependent on interspecific differences in rates of growth. Indeed it would predict the ultimate demise of *O. digyna* at low altitudes, because of its poor ability to fill geometric space. However, matrix models derived from a longer time series of observations would include the second model, because the observations would include rates of patch infill. The value of retaining the two models lies in the separation of the different processes of invader life-cycle characteristics from rates of gap infill and invader growth.

References

Agren, J., Isaksson, L. & Zackrisson, O. (1983). Natural age and size of *Pinus sylvestris* and *Picea abies* on a mire in the inland part of Northern Sweden. *Holarctic Ecology*, **6**, 228–37.

Auclair, A.N. & Cottam, G. (1971). Dynamics of black cherry (*Prunus serotina* Erhr.) in Southern Wisconsin oak forests. *Ecological Monographs*, **41**, 153–77.

Bell, K.L. & Bliss, L.C. (1979). Autecology of *Kobresia bellardii*: why winter snow accumulation limits local distribution. *Ecological Monographs*, **49**, 377–402.

Black, R.A. & Bliss, L.C. (1980). Reproductive ecology of *Picea mariana* (Mill.) Bsl., at tree line near Inuvik, Northwest Territories, Canada. *Ecological Monographs*, **50**, 331–54.

Bray, J.R. & Curtis, J.T. (1957). An ordination of the upland forest communities of southern Wisconsin. *Ecological Monographs*, **27**, 325–49.

Carter, R.N. & Prince, S.D. (1981). Epidemic models used to explain biogeographical limits. *Nature*, **293**, 644–5.

Caswell, H. (1978). Predator mediated coexistence: a non-equilibrium model. *American Naturalist*, **112**, 127–54.

Chapin, F.S. & Chapin, M.C. (1980). Revegetation of an arctic disturbed site by native tundra species. *Journal of Applied Ecology*, **17**, 449–56.

Connor, E.F. & McCoy, E.D. (1979). The statistics of the species-area relationship. *American Naturalist*, **113**, 791–833.

Dahl, E. (1951). On the relation between summer temperatures and the distribution of alpine vascular plants in the lowlands of Fennoscandia. *Oikos*, **3**, 22–52.

Davis, M.B. (1978). Climatic interpretation of pollen in quaternary sediments. In *Biology and Quaternary Environments*, ed. D. Walker & J.C. Guppy, pp. 35–51. Canberra: Australian Academy of Science.

Davis, M.B. (1981). Quaternary history and the stability of forest communities. In *Forest Succession Concepts and Application*, ed. D.C. West, H.H. Shugart & D.B. Botkin, pp. 132–53. New York: Springer-Verlag.

Davison, A.W. (1977). The ecology of *Hordeum murinum* L. III. Some effects of adverse climate. *Journal of Ecology*, **65**, 523–30.

Enright, N. & Ogden, J. (1979). Applications of transition matrix models in forest dynamics. *Araucaria* in Papua, New Guinea and *Nothofagus* in New Zealand. *Australian Journal of Ecology*, **4**, 3–23.

Erdmann, G.G., Godman, R.M. & Oberg, R.R. (1975). Crown release accelerates diameter growth and crown development of yellow birch saplings. *United States Forestry Service Research Paper*, **64**, 104–8.

Evans, G.C. (1972). *The Quantitative Analysis of Plant Growth.* Oxford: Blackwell Scientific Publications.

Forcier, L.K. (1975). Reproductive strategies and the co-occurrence of climax tree species. *Science*, **189**, 808–9.

Gauch, H.G. (1982). *Multivariate Analysis in Community Ecology.* Cambridge University Press.

Gross, K.L. (1980). Colonisation by *Verbascum thapsus* (Mullein) of an old field in Michigan: experiments on the effects of vegetation. *Journal of Ecology*, **68**, 919–28.

Grubb, P.J. (1977). The maintenance of species richness in plant communities: the importance of the regeneration niche. *Biological Reviews*, **52**, 107–45.

Grubb, P.J. (1982). Control of relative abundance in roadside *Arrhenatheretum*: Results of a long-term garden experiment. *Journal of Ecology*, **70**, 845–61.

Hamann, O. (1979). On climate conditions, vegetation types, and leaf size in the Galapagos Islands. *Biotropica*, **11**, 101–22.

Hamann, O. (1981). Plant communities of the Galapagos Islands. *Dansk Botanisk Arkiv*, **34**. Nr 2.

Harper, J.L. (1977). *Population Biology of Plants.* London: Academic Press.

Harper, J.C. & White, J. (1974). The demography of plants. *Annual Review of Ecology and Systematics*, **5**, 419–63.

Hartshorn, G.S. (1975). A matrix model of tree population dynamics. In *Tropical Ecological Systems*, ed. F.B. Golley & E. Medina, pp. 41–51. New York: Springer-Verlag.

Hartshorn, G.S. (1978). Tree falls and tropical forest dynamics. In *Tropical Trees as Living Systems*, ed. P.B. Tomlinson & M.H. Zimmermann, pp. 617–38. Cambridge University Press.

Heinselman, M.L. (1973). Fire in the virgin forests of the Boundary Waters Canoe Area, Minnesota. *Quaternary Research*, **3**, 329–82.

Henry, J.D. & Swan, J.M.A. (1974). Reconstructing forest history from live and dead plant material – an approach to the study of forest succession in southwest New Hampshire. *Ecology*, **55**, 772–83.

Heusser, C.J. (1973). Postglacial vegetation on Umnak Island, Aleutian Islands, Alaska. *Review of Paleobotany and Palynology*, **15**, 277–85.

Heusser, C.J. (1978). Postglacial vegetation on Adak Island, Aleutian Islands, Alaska. *Bulletin of the Torrey Botanical Club*, **105**, 18–23.

Heusser, C.J. (1983). Pollen diagrams from the Shumagin Islands and adjacent Alaska Peninsula, southwestern Alaska. *Boreas*, **12**, 279–95.

Heywood, V.H. (1979) (ed.). *Flowering Plants of the World.* Oxford: Oxford University Press.

Hibbs, D.E. (1982). Gap dynamics in a hemlock – hardwood forest. *Canadian Journal of Forest Research*, **12**, 522–7.

Holt, B.R. (1972). Effect of arrival time on recruitment mortality and reproduction in successional plant populations. *Ecology*, **53**, 668–73.

Hora, B. (1981). (ed.) *The Oxford Encyclopedia of Trees of the World.* Oxford: Oxford University Press.

Hulten, E. (1927–8). *Flora of Kamtchatka and the Adjacent Islands*. Stockholm: Almqvist & Wiksells.
Hulten, E. (1941–50). *Flora of Alaska and Yukon*. Lund: Gleerup.
Hulten, E. (1960). *Flora of the Aleutian Islands and Westernmost Alaska Peninsula with Notes on the Flora of Commander Islands*. Weinheim: Cramer.
Huntley, B. & Birks, H.J.B. (1983). *An Atlas of Past and Present Pollen Maps for Europe: 0 – 13 000 years Ago*. Cambridge University Press.
Jones, N. (1983). Growth studies of selected plant species with well-defined European distributions. PhD thesis, University of Cambridge.
Kanzaki, M. (1984). Regeneration in subalpine coniferous forests. I. Mosaic structure and regeneration process in a *Tsuga diversifolia* forest. *Botanical Magazine*, **97**, 297–311.
Kays, S. & Harper, J.L. (1974). The regulation of plant and tiller density in a grass sward. *Journal of Ecology*, **62**, 97–105.
King, T.S. (1975). Inhibition of seed germination under leaf canopies in *Arenaria serpyllifolia, Veronica arvensis* and *Cerastium holosteoides. New Phytologist*, **75**, 87–90.
Kira, T., Ogawa, H. & Sakazaki, H. (1953). Intraspecific competition among higher plants. I. Competition-density-yield interrelationships in regularly dispersed populations. *Journal of the Institute of Polytechnics, Osaka City University D*, **4**, 1–16.
Kullman, L. (1979). Change and stability in the altitude of the birch tree limit in the southern Swedish Scandes 1915–1975. *Acta phytogeographica suecica*, **65**. 121p.
Kullman, L. (1983). Past and present tree lines of different species in the Handolan Valley. Central Sweden, In *Tree Line Ecology*, Proceedings of the Northern Quebec Tree-Line Conference, ed. P. Morisset & S, Payette, pp. 25–42. Quebec: Centre d'études nordiques de l'Université Laval.
Law, R. (1975). Colonisation and the evolution of life histories in *Poa annua*. PhD thesis, University of Liverpool.
Law, R., Bradshaw, A.D. & Putwain, P.D. (1977). Life history variation in *Poa annua. Evolution*, **31**, 233–46.
Lefkovitch, L.P. (1965). The study of population growth in organisms grouped by stages. *Biometrics*, **21**, 1–18.
Leslie, P.H. (1945). On the use of matrices in certain population mathematics. *Biometrika*, **35**, 183–212.
Lid, J. (1964). The flora of Jan Mayen. *Norsk Polarinstitut Skrifter*, Nr 130.
MacArthur, R. (1965). Patterns of species diversity. *Biological Reviews*, **40**, 510–33.
MacArthur, R. & Wilson, E.O. (1967). *The Theory of Island Biogeography*. Princeton: Princeton University Press.
Mooney, H.A. & Billings, W.D. (1961). Comparative physiological ecology of arctic and alpine populations of *Oxyria digyna. Ecological Monographs*, **31**, 1–29.
Mörner, N.A. (1980). A 10 700 years' paleotemperature record from Gotland and Pleistocene/Holocene boundary events in Sweden. *Boreas*, **9**, 283–7.
Mörner, N.A. & Wallin, B. (1976). A 10 000 year temperature record from Gotland, Sweden. *Palaeogeography, Palaeoclimatology, Palaeoecology*, **21**, 113–38.
Müller. M.J. (1982). *Selected Climatic Data for a Global Set of Standard Stations for Vegetation Science*. The Hague: Junk.
Naka, K. (1982). Community dynamics of evergreen broadleaf forests in southwestern Japan. I. Wind damaged trees and canopy gaps in an evergreen oak forest. *Botanical Magazine*, **95**, 385–99.

Page, C.N. (1979*a*). The diversity of ferns: an ecological perspective. In *The Experimental Biology of Ferns*, ed. A.F. Dyer, pp. 9–56. London: Academic Press.

Page, C.N. (1979*b*). Experimental aspects of fern ecology. In *The Experimental Biology of Ferns*, ed. A.F. Dyer, pp. 551–89. London: Academic Press.

Payette, S. & Lajeunesse, R. (1980). Les combes à neige de la rivière aux Feuilles (Nouveau-Quebec): indicateurs paléoclimatiques holocènes. *Geographie Physique et Quaternaire*, **34**, 209–20.

Peet, R.K. (1984). Twenty-six years of change in a *Pinus strobus*, *Acer saccharum* forest, Lake Itasca, Minnesota. *Bulletin of the Torrey Botanical Club*, **111**, 61–8.

Pigott, C.D. (1981). Nature of seed sterility and natural regeneration of *Tilia cordata* near its northern limit in Finland. *Annales Botanici Fennici*, **18**, 255–63.

Pigott, C.D. & Huntley, J.P. (1978). Factors controlling the distribution of *Tilia cordata* at the northern limits of its geographical range. I. Distribution in north-west England. *New Phytologist*, **81**, 429–41.

Pigott, C.D. & Huntley, J.P. (1980). Factors controlling the distribution of *Tilia cordata* at the northern limit of its geographical range. II. History in north-west England. *New Phytologist*, **84**, 145–64.

Pigott, C.D. & Huntley, J.P. (1981). Factors controlling the distribution of *Tilia cordata* at the northern limit of its geographical range. III. Nature and cause of seed sterility. *New Phytologist*, **87**, 817–39.

Pinero, D., Martinez-Ramos, M. & Sarukhan, J. (1984). A population model of *Astrocaryum mexicanum* and a sensitivity analysis of its finite rate of increase. *Journal of Ecology*, **72**, 977–91.

Preston, F.W. (1960). Time and space and the variation of species. *Ecology*, **41**, 611–27.

Raven, P.H. (1972). Plant species disjunctions: a summary. *Annals of the Missouri Botanic Garden*, **59**, 234–46.

Rejmánek, M. (1976). Centres of species diversity and centres of species diversification. *Evolutionary Biology*, **9**, 393–408.

Richards, P.W. & Evans, G.B. (1972). Biological flora of the British Isles: *Hymenophyllum*. *Journal of Ecology*, **60**, 245–68.

Runkle, J.R. (1982). Patterns of disturbance in some old-growth mesic forests of eastern North America. *Ecology*, **63**, 1533–46.

Salisbury, E.J. (1926). The geographical distribution of plants in relation to climatic factors. *Geographical Journal*, **57**, 312–42.

Salisbury, E.J. (1932). The East Anglian flora: a study in comparative plant geography. *Transactions of the Norfolk and Norwich Naturalists Society*, **13**, 191–263.

Sarukhan, J. & Gadgil, M. (1974). Studies on plant demography: *Ranunculus repens* L., *R. bulbosus* L. and *R. acris* L. III. A mathematical model incorporating multiple modes of reproduction. *Journal of Ecology*, **62**, 921–36.

Shugart, H.H. (1984). *A Theory of Forest Dynamics: The Ecological Implications of Forest Succession Models*. New York: Springer-Verlag.

Smith, H. (1920). Vegetationen och dess utrecklingshistoria i det centralsvenska högfjällsområdet. *Norrlandskt Handbibliotek*, **9**. Uppsala, 238p.

Spurr, S.H. & Barnes, B.V. (1980). *Forest Ecology*, 3rd edn. New York: Wiley.

Symonides, E. (1979). The structure and population dynamics of psammophytes on inland dunes: I. Populations of initial stages. *Ekologia Polska*, **27**, 3–37.

Tateaki, M. (1963). Hultenia. *Journal of the Faculty of Agriculture, Hokkaido University*, **53**, 131–199.

Tatewaki, M. & Kobayashi, Y. (1934). A contribution to the flora of the Aleutian Islands. *Journal of the Faculty of Agriculture, Hokkaido University*, **36**, 1–119.

Tindall, R.W. (1979). Russian America's green legacy to the Aleutians. *Alaska*, **45**, A15-A16.

Toorn, J., van der (1980). On the ecology of *Cotula coronopifolia* L. and *Ranunculus sceleratus* L. I. Geographic distribution, habitat and field observations. *Acta Botanica Neerlandica*, **29**, 385–96.

Toorn, J., van der & Hove, H.J., ten (1980). On the ecology of *Cotula coronopifolia* L. and *Ranunculus sceleratus* L. II. Experiments on germination, seed longevity, and seedling survival. *Acta Oecologia*, **3**, 409–18.

Trimble, G.R. Jnr & Tryon, E.H. (1966). Crown encroachment into openings cut in Appalachian hardwood stands. *Journal of Ecology*, **64**, 104–8.

Turitzin, S.N. (1978). Canopy structure and potential light competition in two adjacent annual plant communities. *Ecology*, **59**, 161–7.

Usher, M.B. (1969). A matrix model for forest management. *Biometrics*, **25**, 309–15.

Usher, M.B. (1972). Redevelopments in the Leslie matrix model. In *Mathematical Models in Ecology*, ed. J.N.R. Jeffers, pp. 29–60. Oxford: Blackwell Scientific Publications.

Wace, N.M. (1961). The vegetation of Gough Island. *Ecological Monographs*, **31**, 337–67.

Wace, N.M. & Dickson, J.H. (1965). The terrestrial botany of the Tristan da Cunha Island. *Philosophical Transactions of the Royal Society, Series B*, **249**, 273–360.

Wardle, P. (1985). New Zealand timberlines 3. A synthesis. *New Zealand Journal of Botany*, **23**, 263–71.

Werff, H., van der (1983). Species number, area and habitat diversity in the Galapagos Islands. *Vegetatio*, **54**, 167–75.

White, J. (1980). Demographic factors in populations of plants. In *Demography and Evolution in Plant Populations*, ed. O.T. Solbrig, pp. 21–48. Oxford: Blackwell Scientific Publications.

White, J. & Harper, J.L. (1970). Correlated changes in plant size and number in plant populations. *Journal of Ecology*, **58**, 467–85.

Whitehead, F.H. (1971). Comparative autecology as a guide to plant distribution. In *The Scientific Management of Animal and Plant Communities for Conservation*, ed. E.O. Duffey, & A.S. Watt, pp. 167–76. *British Ecological Society Symposium*, No. 11. Oxford: Blackwell Scientific Publications.

Whittington, R. (1984). Laying down the −3/2 power law. *Nature*, **311**, 217.

Woodward, F.I. (1975). The climatic control of the altitudinal distribution of *Sedum rosea* (L.) Scop. and *S. telephium* L. II. The analysis of plant growth in controlled environments. *New Phytologist*, **74**, 335–48.

Woodward, F.I. (1983). The significance of interspecific differences in specific leaf area to the growth of selected herbaceous species from different altitudes. *New Phytologist*, **95**, 313–23.

Woodward, F.I. & Jones, N. (1984). Growth studies of selected plant species with well-defined European distributions. I. Field observations and computer simulations on plant life-cycles at two altitudes. *Journal of Ecology*, **72**, 1019–30.

Woodward, F.I. & Pigott, C.D. (1975). The climatic control of the altitudinal distribution of *Sedum rosea* (L.) Scop. and *S. telephium* L. I. Field observations. *New Phytologist*, **74**, 323–34.

Woodward, F.I. & Sheehy, J.H. (1983). *Principles and Measurements in Environmental Biology*. London: Butterworths.

Yaqub, M. (1981). The implications of climate in the control of the distribution of *Verbena officinalis* and *Eupatorium cannabinum*. PhD thesis, University of Wales.

Yoda, K., Kira, T., Ogaa, H. & Hozumi, K. (1963). Intraspecific competition among higher plants. XI. Self-thinning in overcrowded pure stands under cultivated and natural conditions. *Journal of Biology, Osaka City University*, **14**, 107–29.

Young, D.R. & Smith, W.K. (1979). The influence of sunflecks on the temperature and water relations of two subalpine understorey congeners. *Oecologia*, **43**, 195–205.

6

Digest

I have called this principle, by which each slight variation, if useful, is preserved, by the term of Natural Selection.
C. R. Darwin.

The previous five chapters have attempted to investigate, from a range of viewpoints, the principle of the climatic control of plant distribution. The analyses have been in part selected review and in part selected model, in order to stimulate interest and knowledge in this crucial and central theme of ecology. The aims of developing models are at least twofold: to test the predictive efficiency of present knowledge and to provide hypotheses for experimental scrutiny.

The approach is reductionist, in the sense that what are considered to be major themes have been selected at the expense of others which may also be critical in controlling plant distribution but might be classified in the fine detail of the subject. Some of these themes will be briefly discussed in this chapter, both to broaden the general outlook and to continue the theme, instituted in Chapters 4 and 5, of investigation on an increasingly fine scale.

Climate and the distribution of vegetation

The aim of Chapter 4 was to develop a model based on ecophysiological principles to predict the major vegetational zones of the globe. Predictions were based on various plant responses, such as low-temperature survival and evapotranspiration, and on climatic records from meteorological stations which had been averaged over a number of years of observations. This averaging procedure eliminates the natural variability that may be clearly seen for continuous records (Chapter 2). Yet there may often be a clear correlation between climatic variability and average climate. For example, global measurements of annual rainfall (C) and the annual variability (V) of total rainfall are related as:

$$V = 148\ C^{-0.33}, \qquad (1)$$

where variability V is in % and rainfall C is in mm (from Conrad, 1941 and Regal, 1982). Variability therefore increases with decreasing rainfall.

In deserts, where vegetation is strongly controlled by a limited supply of rainfall, plants must also survive years with a less than average rainfall, punctuated by periods with greater than average rain. Ephemeral species will quickly flower and develop in the 'wet' years. Even long-lived and drought tolerant perennial species, such as *Agave deserti*, only appear to be able to establish from seed in these periods (Professor P.S. Nobel, personal communication). In this same desert environment, diurnal and annual fluctuations of temperature may be large, suggesting that high temperature tolerance may be more critical than low temperature tolerance for plant survival. However Nobel & Smith (1983) have suggested for 14 species of *Agave* that their local distributions are controlled not by the ability to tolerate high temperatures, which is high in all cases, but by their ability to tolerate low temperatures.

Even in the warm desert, therefore, minimum temperatures may be more critical to the control of plant distribution than the more obvious feature of drought. Desert vegetation, because of its simple vertical structure and paucity of species, may prove to be ideal for the experimental testing of the hypotheses. Of particular interest might be studies on the climatic control of the distribution of warm and cool deserts, for example. Experiments should aim to investigate mechanisms, rather than settle with correlations, and in addition should consider the control of distribution not only in the poleward direction, where extremes of climate may be most critical, but also in the equatorial direction, where competition may be crucial.

Climate and the distribution of taxa

In order to investigate the climatic control of the distribution of taxa it becomes necessary to split the life of a plant into a number of stages (Chapter 5), each of which is a link in the chain of survival and each of which can dominate the control of distribution. When a stage in the cycle is broken then extinction is likely. This is the approach which has been adopted for the studies of vegetation but with no considerations of which stage of the life-cycle is critical. When one link is weaker than the remainder, survival is still assured, but if the size of the link is a measure of the flux of individuals, then it is possible that a small link, and therefore flux, may also lead to extinction. Models which can predict the outcome of variations in fluxes of individuals from one stage to the next, or which can predict and integrate the influence of variations in physiology on population dynamics (Bazzaz, 1984), must be the way forward to integrate the tremendous range of information on the life-cycles of species (Harper, 1977, Bazzaz, 1979).

The matrix model and the gap-fill model described in Chapter 5 have attempted to satisfy this need. However it is easy to see that the models need modification to consider a variety of other effects which are also under climatic control, such as animal predation (Janzen, 1970; Hubbell, 1980), the patterns of gap fill (e.g. uneven radial infill, or infill within the gap by plant runners or rhizomes (Harper, 1980 & Bell, 1984)), extreme but small scale micro-climatic variations or patterns (Ustin *et al.*, 1984), disease (Day & Monk, 1974) and variations in edaphic features. The effect of climate on the response of plants to soils is readily observed. *Digitalis purpurea*, for example, is only found on soils of low pH in the cool British Isles, but occurs on soils of higher pH in the warmer regions of continental Europe, effects which can be simulated under controlled environment conditions (Whitlock, 1983). As Venable (1984) points out, 'the expression of genetic variation may vary from one environment to another'. The classic transplant experiment where individuals of some taxanomic status are removed from one site and transplanted to other sites, both within and beyond the observed range, has clearly demonstrated genetic variation in response to climate (Clausen, Keck & Hiesey, 1940, 1948) and remains as one of the key experimental approaches to the subject of climate and plant distribution. However the growth of transplants in the field will be in response to many correlated climatic features. So, for example, high wind speed, low temperature and high humidity are generally correlated. The effect of these variables of climate can only be determined in controlled conditions.

The simplest experimental approach would be to grow plants in some 'standard' climate and vary each aspect of climate independently. In the three cases which have been mentioned, this would necessitate three experiments with a range of treatments in each experiment, plus one with all three in combination and perhaps other combinations. Yet the 'standard' conditions will vary through the life of a plant and must also be changed in order to simulate natural conditions. Finally, the life of the plant must be considered, so how long should the experiment run and is it possible to reduce this time by using different aged individuals? The answer to the latter may be unforthcoming because of uncertainty of the nature of the previous climatic history of the individuals and of carry-over effects. Experiments in controlled environments clearly have at least as many conceptual and organisational problems as experiments in the field, and have no promise of achieving any ecologically relevant explanations. Some compromise in which field experiments are backed by controlled environment experiments, with very limited and closely defined objectives, appears to be the only way forward. However, the emphasis should be on carefully designed experiments on vegetation in the field, perhaps with some simple manipulation of climate.

On small spatial scales (of less than 1 m^2) Snaydon (1970), Antonovics (1976), Davies & Snaydon (1976) and Fowler & Antonovics (1981) have shown, by transplant experiments, that significant genetic variation occurs in life-cycle characteristics between individuals within a population of one species. This variation arises as a result of the local selection pressures associated with a number of features such as soil and micro-climate characteristics and plant competition. The time scale of this population divergence may be short. Snaydon & Davies (1982), for example, record a period of 6 years for the genetic divergence of populations of *Anthoxanthum odoratum* in response to a change in soil chemistry and competitive interactions.

In contrast, investigations by Palmer (1972) on *Trifolium arvense*, Dunbier (1972) on *Medicago lupulina* and Guries & Ledig (1982) on *Pinus rigida*, found no evidence for significant population divergence in a range of climatic conditions. It therefore follows that any studies on plant species should consider genetic variation within the species, both within and at the margins of its range (Solbrig, 1980), and on the fine scale within local populations. As Turkington & Aarssen (1984) note, when speaking of competition, the crucial contest is 'not species pitted against species', but rather 'genotype against genotype'. Logical and exciting this may be, but it makes no concession to the sheer volume of investigations that are required at all levels from the individual to the biome. Nevertheless, in the search for both precision and predictability in ecology, the reductionist approach is considered to be the best way forward (Harper, 1982).

Yet, new approaches arrive and provide unexpected and comprehensive approaches to intractable problems. In 1958, Clausen and Hiesey published a monograph on the genetic structure of *Potentilla glandulosa*. They investigated the genetic constitution of three sub-species of *P. glandulosa*, ssp. *nevadensis* (subalpine), ssp. *typica* (sea-level) and ssp. *reflexa* (foothill). They crossed these subspecies and estimated from the F2 or F3 generations the minimum numbers of pairs of alleles which governed the inheritance of 19 plant characters. Briggs & Walters (1984) point out that this technique has its limitations. In spite of these problems there was clear evidence for considerable genetic difference between the subspecies, although some features with obvious ecological relevance, such as winter dormancy, frost susceptibility and crown height, were estimated to be controlled by only 3 – 5 pairs of alleles at unlinked gene loci. Can we look forward to a time when biological engineering will be sufficiently developed and available that we can change, in a predictable manner, such crucial genetically determined responses as these? Can we also modify the genotype so that we have a marker to follow these individuals, in the same way that Barber (1955, 1965) took advantage of a simple but readily observable polymorphism (glaucous or green leaves) in his studies on the response of *Eucalyptus* to altitude? This could certainly provide a whole range of traceable plant markers to order and for investigating the effects of climate on plant distribution.

References

Antonovics, J. (1976). The population genetics of mixtures. In *Plant Relations in Pastures*, ed. J.R. Wilson, pp. 233–52. Melbourne: CSIRO.

Barber, H.N. (1955). Adaptive gene substitution in *Eucalyptus*. *Evolution*, **9**, 1–14.

Barber, H.N. (1965). Selection in natural populations. *Heredity*, **20**, 551–72.

Bazzaz, F.A. (1979). The physiological ecology of plant succession. *Annual Review of Ecology and Systematics*, **10**, 351–71.

Bazzaz, F.A. (1984). Demographic consequences of plant physiological traits: some case studies. In *Perspectives on Plant Population Ecology*, ed. R. Dirzo & J. Sarukhan, pp. 324–46. Sunderland, Massachussetts: Sinauer Assoc. Inc.

Bell, A.D. (1984). Dynamic morphology: a contribution to plant population ecology. In *Perspectives on Plant Population Ecology*, ed. R. Dirzo & J. Sarukhan, pp. 48–65. Sunderland, Massachussetts: Sinauer Assoc. Inc.

Briggs, D. & Walters, S.M. (1984). *Plant Variation and Evolution*. Cambridge University Press.

Clausen, J. & Hiesey, W.M. (1958). Experimental studies on the nature of species. IV. Genetic structure of ecological races. *Carnegie Institution of Washington, Publication No. 615*.

Clausen, J., Keck, D.D. & Hiesey, W.M. (1940). Experimental studies on the nature of species. I. The effect of varied environments on western American plants. *Carnegie Institution of Washington, Publication No. 520*.

Clausen, J., Keck, D.D. & Hiesey, W.M. (1948). Experimental studies on the nature of species. III. Environmental responses of climatic races of *Achillea*. *Carnegie Institute of Washington, Publication No. 581*.

Conrad, V. (1941). The variability of precipitation. *Monthly Weather Review*, **69**, 5–11.

Davies, M.S. & Snaydon, R.W. (1976). Rapid population differentiation in a mosaic environment. I. The response of *Anthoxanthum odoratum* populations to soils. *Evolution*, **24**, 257–69.

Day, P. & Monk, C.D. (1974). Vegetation patterns on a southern Appalachian watershed. *Ecology*, **55**, 1064–74.

Dunbier, M.W. (1972). Genetic variability in *Medicago lupulina* L. across a valley in the Weka Pass, New Zealand. *New Zealand Journal of Botany*, **10**, 48–58.

Fowler, N.L. & Antonovics, J. (1981). Small-scale variability in the demography of transplants of two herbaceous species. *Ecology*, **62**, 1450–7.

Guries, R.P. & Ledig, F.T. (1982). Genetic diversity and population structure in Pitch Pine (*Pinus rigida* Mill.). *Evolution*, **36**, 387–402.

Harper, J.L. (1977). *Population Biology of Plants*. London: Academic Press.

Harper, J.L. (1980). Plant demography and ecological theory. *Oikos*, **35**, 244–53.

Harper, J.L. (1982). After description. In *The Plant Community as a Working Mechanism*, ed. E.I. Newman, pp. 11–25. Oxford: Blackwell Scientific Publications.

Hubbell, S.P. (1980). Seed predation and the coexistence of tree species in tropical forests. *Oikos*, **35**, 214–29.

Janzen, D.H. (1970). Herbivores and the number of tree species in tropical forests. *American Naturalist*, **104**, 501–28.

Nobel, P.S. & Smith, S.D. (1983). High and low temperature tolerances and their relationships to distribution of agaves. *Plant, Cell and Environment*, **6**, 711–9.

Palmer, T.P. (1972). Variation in flowering time among and within populations of *Trifolium arvense* L. in New Zealand. *New Zealand Journal of Botany*, **10**, 59–68.

Regal, P.J. (1982). Pollination by wind and animals: ecology of geographic patterns. *Annual Review of Ecology and Systematics*, **13**, 497–524.

Snaydon, R.W. (1970). Rapid population differentiation in a mosaic environment. I. The response of *Anthoxanthum odoratum* populations to soils. *Evolution*, **24**, 257–69.

Snaydon, R.W. & Davies, M.S. (1982). Rapid divergence of plant populations in response to recent changes in soil conditions. *Evolution*, **30**, 289–97.

Solbrig, O.T. (1980). Genetic structure of plant populations. In *Demography and Evolution in Plant Populations*, ed O.T. Solbrig, pp. 49–65. Oxford: Blackwell Scientific Publications.

Turkington, R. & Aarssen, L.W. (1984). Local-scale differentiation as a result of competitive interactions. In *Perspectives on Plant Population Ecology*, ed. R. Dirzo & J. Sarukhan, pp. 107–27. Sunderland, Massachusetts: Sinauer Assoc. Inc.

Ustin, S.L., Woodward, R.A., Barbour, M.G. & Hatfield, J.C. (1984). Relationships between sunfleck dynamics and red fir seedling distribution. *Ecology*, **65**, 1420–8.

Venable, D.L. (1984). Using intraspecific variation to study the ecological significance and evolution of plant life-histories. In *Perspectives on Plant Population Ecology*, ed. R. Dirzo & J. Sarukhan, pp. 166–87. Sunderland, Massachusetts: Sinauer Assoc. Inc.

Whitlock, R.J. (1983). Climatic and edaphic factors as determinants of the geographic distribution of *Inula conyza* and *Digitalis purpurea*. PhD thesis, University of Cambridge.

Index